Preface to Third Edition.

Since the first publication of this work in 1905, two large editions have been disposed of, and another is required. This is especially gratifying, in view of the fact, that the sale of the book has not been urged, except as one purchaser or patient would recommend it to another.

The succeeding editions have not differed materially, in describing the method of treatment, except in the effort to make the directions more explicit and to provide for certain contingencies.

The fact that I have a large correspondence with people who contemplate taking, or have, or are taking a milk diet in some way, has afforded me every year valuable information as to their needs, and enabled me to emphasize, or make clearer, features which may not have been sufficiently plain for everyone in the earlier editions.

Valuable contributions regarding the

treatment are frequently offered me, and, after a thorough test, may be incorporated in the book.

Unless otherwise credited, all the statements made regarding reactions, results and the method of operation of the treatment, are from my own observation.

Orders for the book have come from every part of the United States, from Canada, England, and other foreign countries; translations, in whole, or in part, have also been made into other languages.

I hope the new book will meet with the same favorable unanimity that its predecessors received.

C. S. P.

Burnett, California
January 31, 1911

Milk Diet as a Remedy for Chronic Disease

by

Charles Sanford Porter

APPLEWOOD BOOKS
Bedford, Massachusetts

Milk Diet as a Remedy for Chronic Disease

was originally published in

1911

ISBN: 978-1-4290-1171-6

Prepared for publishing by HP

MILK DIET

AS A REMEDY

For Chronic Disease

BY

CHARLES SANFORD PORTER, M. D.

THIRD EDITION

BURNETT P. O., CALIFORNIA
(City of Long Beach)
1911

Table of Contents.

Disease can only be cured by and through the blood, and its circulation in the living parts of the body.

Any means that improves the quality, the quantity, and the movement of the blood, will assist in eliminating disease.

The treatment described in this book makes the purest and richest blood possible, and ALWAYS increases the amount and rapidity of the circulation.

SYNOPSIS

Briefly stated the milk cure consists of the following factors:

First *Complete rest for all the organs of the body except those concerned in the production and circulation of the blood, and those connected with the elimination of waste and poisonous matter.*

Second *An ample supply of the only food that will make an immediate large production of blood possible,—milk.*

Third *An unlimited quantity of pure air to oxidize and cool the blood, and carry off the expired gases.*

Fourth *Warm water baths, to soften the skin, equalize the circulation, and regulate the body heat—and, last but not least,*

Fifth *When the body is ready for it, EXERCISE, to strengthen the muscles, expand the lungs, limber the joints, stimulate the circulation, increase the elimination, purify the blood, develop normal secretions, train the nerves, and, generally, to fix and make permanent the benefits acquired while resting and building up the body.*

PRESS PRINTING COMPANY

LONG BEACH, CALIF.

INTRODUCTION

Milk, and milk products, have been used and highly esteemed as food by all nations possessing mammalian animals since the earliest records of history. At the present time some of the older European countries are consuming two or three times as much milk and cheese per capita, as the United States does.

A good food is a good remedy, and, as disease is only a disturbance of the mechanism of nutrition, it is only natural that the use of milk in ill health should be almost as old as its use as a food in health.

Hippocrates advised consumptives to drink large quantities of asses' milk. Camel's milk, and whey cures were practiced by the Arabian physicians. Homer called the Scythians Galactophagi, or feeders on milk, and Herodotus describes their methods of handling mare's milk.

In recent times the popularization of the milk cure has been largely due to the efforts of

Russian and German physicians. Karrick, Karel, F. von Niemeyer, Winternitz and Bremer have advocated it in the treatment of many chronic diseases. In the United States, Donkin, in 1868 called the attention of the medical profession to the fact that diabetes could be cured by large quantities of milk, and said that some cases could take as much as fourteen pints per day. Weir Mitchell, who has had, perhaps, the greatest experience of any American physician with chronic disease, says: "It is difficult to treat any of these cases without a resort at some time more or less to the use of milk."

The use of milk in the diet for nearly all cases of chronic disease, is advised by all the principal textbooks with which I am familiar. The mistake is made by many of them of combining other foods, even meats and eggs, with milk, and, consequently they usually add the proviso "if it agrees." Very few physicians understand the proper amounts of milk to be given, and the proper way to give it in order to assure assimilation. The method of preparing the patient for the milk diet, and his conduct while taking it, had never been published, that I am aware of, previous to the printing of the first edition of this work, in 1905.

My first introduction to the possibilities of an exclusive milk diet was in 1884 while living in New York City. A friend was cured completely and permanently of a rather serious condition, by following the advice of a gentleman recently from Germany and familiar with the milk cures practiced there at that time. Other mutual friends adopted the same plan for the relief of various ills, and all with good results. Shortly afterward I took the "milk cure," as we understood it then, for a condition bordering on nervous prostration. I not only overcame that condition, but was cured of hay fever which had claimed me as an annual victim for seven or eight years, but has never since returned.

From that time to the present, I have never ceased to advocate the milk cure, and, of the thousands of cases of chronic disease that have taken the treatment in the manner I recommended, I do not know of one person who has regretted it; I do not know of one that was not benefited.

Such remarkable results have, of course, resulted in extending the treatment throughout the country. Many physicians, and some sanitariums, have endeavored to use the method. I regret that many, if not all of these, have made changes in the original plan, which was apparently too simple. Very many times doctors and other patients under my care have advised me to add something to the milk, or to the method of giving it, in order to make the process more mysterious, and more attractive to many people, and more lucrative to myself. They said that people who had suffered for years without relief, traveling to many health resorts, and to noted specialists, taking expensive and elaborate treatments, would not easily be induced to use a method apparently so simple as to be within the reach of almost every person, or household, without expert advice.

While admitting the force of the argument, I have always replied that there was only one way to do the milk cure, and that I would go on to the end advocating that way. For twenty-seven years I have watched the results of this way, always willing to add anything of real benefit, that would not interfere with the results we were already getting, always investigating methods of treatment that seemed to have merit, or that made claims to be able to do more, or even as much as we could, but, with the exception of the preparation of the patient for the milk diet, and the method of finishing the treatment, it remains practically the same as when first introduced to this country.

It is true that my experience has shown me what kind of milk gives the best average results, and what parts of the treatment may be omitted or modified in certain cases, and I may also say that I have spent a great deal of time investigating the why and wherefore, the reason for doing certain things, and not doing others, and the cause of certain symptoms and results that occur during a course of milk diet.

This book is particularly intended for patients taking the treatment,—**and those who are under my personal care, or corresponding with me, must be perfectly familiar with it.** For this reason it is made as brief and simple as possible, but it contains all necessary instructions regarding the treatment. It must be read carefully before beginning the treatment, and again soon after commencing, unless the patients' condition is such that reading is unadvisable, when it should be read to them, if possible.

The book is not large; the size has been kept down so that almost any invalid can hold the book, and read it, without too much physical or mental effort.

The present edition has taken a long time to prepare, because it had to be written between the almost constant calls of my daily work, and also because the contents comprise, almost entirely, my personal experience. My library on Milk is as complete as I can make it, but in the scores of books, pamphlets, and periodicals on the subject, there is little to draw on that would be of practical use in this work.

CHAPTER I.

PRELIMINARY ARRANGEMENTS.

Before commencing a course of milk diet, certain preparations are necessary. These preliminaries must be arranged beforehand, because the treatment always includes complete rest, for a time, at least.

The consideration of the apartment where the patient is to remain is of first importance. It must be remembered that, no matter what the previous habits of the patient may have been in this regard, a very large supply of fresh air will be required, if not at first, within about forty-eight hours. A room may be used, and often is, but the best results, in my experience, have followed the use of outdoor bedrooms, such as pavilions, screened porches, roofs, sheds, lean-tos, or even a good bed with nothing over it. In most climates some protection is required from the rain, snow, sun, or wind. On the whole, perhaps there is nothing more satisfactory than a pavilion, partially boarded or latticed up on the sides, with a good water-tight roof, and insect-proof screen over the openings all around. Some of the openings should extend to the roof, or ceiling, and some of them should come down to the floor. It is the lack of these that prevents a room being equal to an outdoor place; no matter how many windows there may be, there is a dead space above the tops of the windows where warm air accumulates, and there is a space between the bottoms of the windows and the floor where the heavy gases, such as carbonic acid, lie more or less stagnant, until stirred up by some breeze of unusual strength or direction. Dust is also deposited in these dead air spaces. Anyone who has not tried living and sleeping in a space open from floor to roof, even on only one side, cannot realize what a constant difference there is between the air in such a place and the air in a room, no matter how well ventilated it may seem to be.

A room is always more or less draughty, with the windows open, while in these outside places the circulation of air, while thorough, is almost imperceptible, so gently and easily is the change made.

It is the retaining of the gases and other cast-off material from the body in the room that makes indoor life so much more unhealthy, compared with life in the open. The greater warmth, too, indoors, prevents the same degree of oxidation that is possible outdoors. The cooler the air, the purer it is, as a rule, and the more oxygen we are able to absorb. There are probably other substances besides simple oxygen, in fresh air, that are necessary to our wellbeing.

When you have decided upon a suitable location to stay in while taking the milk, arrange for a comfortable bed, preferably one with a hair mattress. A hard bed, or a bumpy one, becomes irksome before the skin has developed the protecting pad of flesh that belongs over the bony points. The head of the bed should be toward the openings where the light and air enter. Do not make the common mistake of putting the feet out in the center space, in a current of air, and the head in some corner where the circulation is at a minimum. The reverse should be the rule.

Beds with solid headboards or footboards should not be used. Procure an iron bed, or a couch or cot without any headboard. Of all things do not attempt to sleep in a modern folding bed where the head is put in a box-like space, eminently more suited to the destruction of one's health than to its restoration.

The bed clothes should be woolen blankets by preference, with linen sheets, fastened at the foot, and folding down from the head of the bed, so that the patient can easily turn down a fold or two when less covering is required.

In certain cases where there is much perspiration, or exhalation from the body, it is a wise plan to use a set of bed linen not over twenty-four hours at a time, not necessarily increasing the laundry expense, but putting the change of linen to air while the other set is in use. Remember that it is necessary to stay in bed all the time, except when bathing, or performing other necessary acts, and that the skin is an important breathing organ, and must not be surrounded by foul odors.

The sleeping garments should be changed twice a day, morning and night. I think a gown is preferable to pajamas, because it is very important that there be no constriction around the waist. Garments requiring to be buttoned, or belted, around the waist, interfere with the proper development of the organs contained in the abdomen, and also prevent, to some extent, abdominal breathing.

I am explicit about these directions, because a very rapid growth and development will take place in the organs of the digestive system, the stomach, liver, intestines, pancreas, etc., and this growth is greater in the first week than during any subsequent period. It is during this first week that the success or failure of the milk cure is usually determined, and this growth, or development, MUST not be interfered with.

If possible, the patient should be within easy reach of the toilet and bathroom. There must be no dressing to go outside the room to a toilet. Have a capacious slopjar in the room and a urinal to use in the bed, especially in cold weather. By having the jar near the bed, the urinal can be used, and emptied into the jar, without getting up, or exposing the person.

A small table, or stand, about two feet high, is required near the head of the bed, to set the milk can and glass on, and for such other small articles as may be required.

A two-quart tin can, or measure, is the most convenient and best receptacle to keep the milk in at the bedside. It is lighter than any pitcher, and unbreakable. Have two napkins to cover the milk can and glass between drinks. Two glasses will be needed, marked in some manner to indicate 5, 6, or 7 ounces of milk. A ring can be scratched around a plain glass with a file at the proper point.

An old established custom in the milk cure is that of using one glass for twenty-four hours, without washing. I still adhere to this plan in my personal practice, unless patients find it disagreeable, or the weather is very warm. In the latter case, it is necessary to serve a clean glass with nearly every quart of milk, or the residue remaining after drinking will sour the next glassful of milk.

Doubtless many patients could use a clean glass every time they took a drink, and still derive the greatest benefit, but it is possible, if not probable, that certain cases would find it unwise to change their glass oftener than once or twice daily. I hesitate to change methods that have proved efficacious for so many years.

Of the many places which have started

up in the last twenty-five years with the intention of giving an "improved" milk diet treatment, I do not know of any now in existence, except one or two recent ones.

A clock must be located where it can easily be seen from the bed. Clocks striking the hours and half hours are a great aid in calling the patient's attention to drinking time. Good clocks of this description can be purchased from $2 up.

Outside of the necessary articles mentioned, the less furniture there is the better it will be. Chairs for visitors are not particularly required, for there should be no visitors. If absolutely necessary, visits may be tolerated, but never for longer than half an hour at a time.

A daily warm water bath will be required and the arrangement of the bathing facilities is one of the things that requires careful attention. It is necessary for the patient to enter the tub while the water is somewhat cooler than the body, and then gradually warm the bath to the body temperature, or to such a temperature as will be entirely comfortable. This necessitates a reserve supply of hot water, which may be drawn on at intervals during the bath as the water cools off.

The ordinary 30-gallon reservoir, used in connection with a range in most households, is not often satisfactory, because drawing the necessary amount of hot water to prepare the bath leaves no surplus, and it is most annoying to open the hot water faucet and get cold water.

However, if the tank is full of hot water, and the fire in the stove is kept going it may work all right, but there **must** be hot water up to the end of the bath. The instantaneous gas heaters, if properly arranged, are satisfactory. If the heater is in the bathroom, it must have a flue carrying the fumes outside of the room. The best arrangement is to have a gas heater in connection with a reservoir, preferably in another room, so that the hot water when not being drawn into the tub, will be collecting in the reservoir. I have found the "Reliable" heater, made in Cleveland, perfectly satisfactory. Those instantaneous heaters discharging the hot water directly from the heater into the tub are not so well adapted to the purpose, and some of them are termed by plumbers "contaminating," that is, the gas fumes come in contact with the water, or pass through it as they escape. Water of this character is dangerous to take long baths in.

There is serious objection to having the water heating apparatus in the bathroom, unless the room is large and well ventilated. The heater uses up more oxygen than the lungs of several people would. Many fatalities have occurred in Southern California from instantaneous heaters causing the asphyxiation of the inmates of bathrooms, perhaps chiefly on account of the habit some people have of shutting the bathroom up tightly while bathing.

The tub itself is a matter of considerable importance. I have not yet seen a modern white enameled iron tub that seemed as satisfactory as the old copper tubs, chiefly on account of the shape. The iron tubs are moulded somewhat like a huge box, with flat bottom and vertical sides. Even the head of the tub where the bather's head and shoulders rest, goes almost straight down, whereas the old style had a gentle slope about two and a half feet long, making a comfortable support for the upper part of the trunk and head. The copper tubs had a rounding bottom which fitted the body better, and did not require so much water to cover one, and the metal itself being thin, was rapidly warmed by the hot water, while the thick iron tubs now used require the expenditure of considerable heat simply to warm up the tub. The iron tubs stand up so high as to be difficult for a weak person to enter, and serious accidents have occurred on account of the bather slipping as he left the tub. Another objection is the location of the overflow so near the bottom that the tub will only hold a few inches of water. This latter fault may sometimes be remedied by unscrewing the fixture and covering the outlet with a thin rubber sheet, or filling it up with putty. Sometimes the overflow may be stopped by simply putting a piece of paper over it, when the force of the water will hold the paper tight against it. The tub ought to be deep enough and long enough to hold sufficient water to cover the shoulders when the patient is extended at full length, and for this purpose a six-foot tub is usually necessary. A five and a half foot, or even five foot tub may be used by short people, or ladies, but the six foot is best. A canvas head rest may be used, if necessary, or a rubber cushion, or hot water bag full of air, to rest the head on. The trouble with most ladies is that they object to wetting the hair, while men as a rule, enjoy lying in the tub with the water up to their mouths, and it is best that they should do this.

The patient should have a bathrobe to wear in going from the sleeping room to the bathroom, and a pair of easy slippers. Felt slippers are the best, as they do not require stockings, and are warm and comfortable. Hundreds of times I have seen patients, after taking a warm bath, leave the bathroom with only bathrobes and slippers on, go outdoors to their beds, in all kinds of weather, and I never knew of any of them "taking cold."

In regard to milk, a few necessary general rules will be given here. What is required is good, clean milk as it comes from the cow, without the removal or addition of any substance whatsoever. Boiled, sterilized or pasteurized milk, or milk artificially preserved in any way, cannot be used for this treatment.

In well-managed modern dairies the handling of milk is so systematized that there is no particular trouble in keeping the milk sweet until used. Dairies that are not cleanly, or have not proper appliances, often use some means of preserving the milk, by stopping the activity of the acid-forming bacteria. These bacteria are not dangerous to health, and the methods of restraining or destroying them are without effect on the bacteria of consumption, typhoid, or other fevers that might contaminate milk in certain places. Prolonged boiling will destroy any germ, but boiled milk alone will not sustain life in either the infant or the adult. Pasteurizing milk or heating to 150 F., or less, can have no effect on the pathogenic bacteria and renders it unsuitable for human use. Dogs fed on pasteurized milk only, are liable to have the mange and other disorders, while others of the same litter thrive on sweet and sour milk.

There are several chemical preservatives sold to dairymen by manufacturers who claim they are harmless. They are prohibited by the laws of most states. Some of them containing borax are not exactly poisonous in the amount one would ordinarily get in milk; but they render the milk much less digestible and in a weak baby or invalid adult might readily be the contributing cause of death. Others like salicylic acid, or formaldehyde or formalin, are distinct poisons. There is no harmless preservative of milk; whatever prevents its decomposition will also render it more or less indigestible.

The manner in which milk is handled makes a great difference in its keeping qualities. Milk which is cooled and aerated immediately after being drawn, will keep for days; while, on the other hand, milk which is left to stand with the animal heat in it, will often be stale within twelve hours, and sour in less than twenty-four hours.

Milk from Holstein cows is the best for the purpose, next that from Durhams or Shorthorns, and last that of the pure Jersey and Guernsey, or Alderneys, as the two latter breeds were formerly called. Milk from Jersey cows may be used, but it should usually be skimmed after standing two to four hours to reduce the amount of cream. Many people anxious to gain weight, think they should take all the cream possible. This is a mistake, as the fat in the milk does not normally make flesh in the body. The flesh built up on a milk diet is derived almost entirely from the proteids and carbohydrates, namely: casein, albumin, etc., and milk sugar. If the fat of a full milk diet was deposited in the body it would mean a gain of about half a pound of pure fat daily.

Milk contains all the salts necessary for the building up of every part of the body. It has iron, potassium, phosphorous, sodium, lime, magnesium, fluorin, etc., and altogether contains about twenty elements.

Dairy milk, or milk from a herd of cows gives a more even average of fat and other contents than the milk from one cow would. There is no advantage in having one cow set aside for your use, unless by so doing you secure Holstein milk.

The milk should be delivered fresh, morning and evening, about two-thirds of the total quantity in the morning, and one-third at night. In the cities the milk is usually ten to fourteen hours old before being delivered. Many of my patients have taken the diet successfully under these conditions, but I think the average results are better with fresher milk.

CHAPTER II.

STARTING THE TREATMENT.

In severe cases of illness, the success of the milk cure depends on the faithfulness with which the details are followed. Some of these details often seem unimportant to those who know little of the treatment, but, in any case where a successful result has not been obtained, it has always been easy to point to faults of commission or omission.

It is true that many people have derived great benefit from a milk diet taken otherwise than as I advise, or only partially following my instructions, but I believe that the plan I give herein is one that is always successful, enabling the patient to take the proper amount of milk, and secure the desired results, without any danger.

Before commencing the milk diet, it is usually advisable, and often necessary, to take a fast from ordinary foods.

For the ordinary case, where the digestion

is more or less impaired, and particularly where constipation is present, the fast should continue at least 36 hours, but the patient is allowed to eat ripe fresh and dried fruits, in such quantities as may be eaten with a relish, and as much water may be taken as possible with comfort.

While I have often started patients on milk only five or six hours after their last meal, sometimes I have regretted it and thought that a day's fast would have saved time. If there is a class of patients who can do without the fast, it is the thin, weak, anemic people, such as consumptives, neurasthenics, etc., especially those whose bowels are in the habit of moving freely every day. Such patients take milk greedily; they soak it up like a sponge, there is no initial constipation nor nausea, and the rapid increase in circulation causes a quick elimination of the impurities in the blood and bowels.

On the other hand, those who are stout, plethoric, rheumatic, gouty, dropsical, constipated, or who have had skin or blood disease, diabetes, headaches, coated tongue, prolapsed or dilated stomach, or any displaced organ, should take at least one day's fast, and many people will be benefited, and gain time in the end, by extending the fast over several days.

Those who are not accustomed to fasting periods are usually agreeably surprised to find there is no particular inconvenience to this part of the program, and when the time comes to start in drinking milk, it goes down with a relish; the stomach makes no objection, and the bowels move naturally. Another important consideration is the fact that the organs of digestion are, so to speak, caught at low tide, at their minimum bulk and activity, and building up rapidly in size and function, as they do on the milk diet, while the mind and body are in a state of as complete rest as possible, there is a natural tendency to make good cells, good tissues, and healthy organs, and to overcome any abnormal habit or loss of natural function that may have been contracted by any organ.

During the fast it is not necessary to take rest, or refrain from the usual work or habits; in fact I think most patients are benefited by active exercise the day before commencing the milk.

In beginning the diet, take the weight in the morning, with as little clothing on as possible. Make a list of what you wear, and at the end of the week, wearing the same outfit, weigh at the same hour of the day. More exact results are secured by emptying the bladder each time before weighing.

Measurements should be taken of the limbs, hips, waist, neck, and especially of the chest, both expanded and contracted, and comparisons made from time to time.

In regard to the amount of milk to be taken, I will make the following statement: The average adult, when consuming daily two to four quarts of milk containing 4 per cent. of butter fat and 9 per cent. of other solids, will not lose flesh; with another quart or two they will gain weight, and with a still further increase of a pint or two they will secure the necessary energy and stimulation to throw off disease.

There are several arbitrary rules as to the quantity necessary, such as taking an ounce and a half of milk for each pound of the normal weight or the highest weight in health, or, taking a quart of milk for every foot in height, but none of these will fit all cases. They are, however, a useful guide, especially in estimating the amount to be given children, when the first rule can be safely followed.

The last thousand cases that I have had under observation have averaged about six quarts of milk daily, containing from 3½ to 4 per cent. of butter fat, and 9 per cent. of solids not fat. The males usually go over that amount, and the average female patient will take slightly less.

It is wrong, if not positively dangerous, to attempt the exclusive milk diet on any amount of milk less than that required to noticeably stimulate the circulation and promote body growth. A possible exception might be made in the case of convalescents from severe, acute fevers, where a few glasses of milk daily might keep them going temporarily until the ability to digest solid food was recovered. Even in those cases, water would be a safer drink, and probably would do as much good.

There is no half-way method of taking the milk diet for people who have much the matter with them. Enough milk must be taken to create new circulation, new cells, and new tissue growth, and cause prompt elimination of the waste and dead matter that may be poisoning the system.

A patient should start with the full amount

of milk; cases that begin on a smaller quantity and try to work up to the proper amount, often fail to get the best results. They get the stomach in the notion of taking three or four quarts, and then find it difficult to increase the amount, while those who start on, say, six quarts daily, have little or no trouble after the first day or two. In beginning in this way we take the stomach by surprise, and as the milk keeps coming, the stomach is compelled to dispose of it, and soon does so, in the natural way, without difficulty. It is rather common for patients to say, the afternoon of the first day, that they feel so full they cannot take another glass, but as they continue taking the regular quantity, ways and means are provided, and the sensation disappears the same day or that night, and does not return.

On the contrary, if you humor the stomach, and stop when it desires you to, you will likely have to repeat the whole process.

It must be remembered that stomachs of this kind are not normal, and have been out of condition so long that they are not competent judges of what is best for them.

On the morning the milk diet is commenced, the patient remains in bed and takes the first drink as soon as the milk is available, but starting on the even hour, or half hour, and takes the same amount every half hour. The next, and following day's drinking begins as soon as the patient is awake in the morning, using the milk supplied the previous evening. The amount of milk taken in twenty-four hours is calculated from the time the first glass of the new day's supply is taken, until the same time the next morning.

If six quarts is the daily amount, use a glass marked to contain six ounces; if seven quarts is the allowance, take seven ounces in a drink. If five and a half quarts are taken, the glass should hold five and a half ounces, and so on. Using these amounts there will be 32 drinks taken in twenty-four hours. If the first drink is taken at 6 a. m., and none are missed, by 8:30 p. m. 30 drinks will have been taken, two to be taken anytime in the night when awake.

This is the only way that such an amount of milk can be absorbed by a weak stomach, and it IS always absorbed, digested, or discharged, where the directions here given are followed.

It is necessary to be exact as to the time and quantity taken. Each glass should be sipped slowly, taking several minutes to finish it. The milk must be mixed with the secretions of the mouth. Do not gulp it, or let it run down the throat, as you might water. Now and then I come across a patient who will take long draughts of milk, say two ounces at a pull, but drawn into the mouth in a rather small stream. They are young people with active salivary glands, and doubtless the action of sucking the milk through a small mouth opening, at the same time draws saliva into the mouth. Such patients say the milk tastes better to them taken in that manner than it does when taken in small swallows and "swished" around in the mouth by the tongue, but the latter is the safest way to start on. A straw, or glass tube, or drinking cup may be used.

Many patients sleep more than half the time. If asleep when drinking time comes, take your glass when you awaken, but do not try to make up for lost time. Continue thirty minutes apart. Milk is supposed to require about one and a half hours for digestion, and all dietetic plans before this have allowed at least that much time between meals. I use the half-hour interval because it gives the best results. Milk is curdled as soon as it arrives in the stomach; the salts and water begin to be absorbed immediately, other portions are passed on to the intestines, where the fat is quickly absorbed by the lacteals. The nitrogenous portions may not be taken up into the blood for twenty-four hours. So it is useless to set any particular time for the digestion of milk or other food. Doubtless a part of the milk will still be in the stomach at the end of thirty minutes, but its mixture with a fresh portion has no bad effect. On the contrary it works well in practice.

A patient, in describing the effect, once said: "After fairly started the first glasses seem to pull the others after them by suction."

If an invalid's stomach is very weak, or particularly deficient in the digestive juices, and especially if the milk is taken too rapidly, tough curds may be formed in the stomach, which are slow and hard to digest. In the vomit of persons who were drinking quantities of milk too quickly, or at too low a temperature, I have seen these cheesy bodies so large and firm that it seemed impossible that they could have come up through the œsophagus. Where the conditions that I recommend as to rest, bathing, air, and the small frequent and regular drinks of milk have been followed, I have never known of these curds being formed in such amounts as to prevent their digestion, with the exception of a few very weak people who were attempting to take their milk cold, or at the temperature of the room.

I usually start patients on milk which is near the room temperature, or at least not below 60 degrees F., but if there are symptoms within the first two or three days of indigestion, distress in stomach, nausea, or vomiting of thick curds, the patient goes on warm milk immediately and does not take any cold milk for two or three weeks.

In cold weather, if any trouble of this kind is anticipated, it is better to start on warm milk at the beginning, but in nineteen cases out of twenty, especially in mind weather, it is unnecessary to warm the milk.

On cold nights, if drinking the cold milk prevents the patient getting to sleep again, provision should be made for warming it.

There is only one satisfactory way of warming the milk, and that is to have a pan containing about three inches of warm, almost hot, water, and set each glass of milk in it for two or three minutes until it is warmed through. The milk ought to be about blood heat, although it can be heated in this way to 120 degrees without harming it, but the milk must not be left long in the water, and must be taken immediately. No more than one glass can be heated at a time. The most convenient way of keeping the pan of water hot is a small oil stove, kept burning continuously at such a heat as may be necessary. If the patient has no nurse the whole arrangement must be set near the bed so the patient can prepare the milk without getting up. Taking this warmed milk, according to the other directions, indigestible curds are never formed.

The secretions of the mouth may have no digestive action on milk, as there is no starch in the latter, but the mechanical effect of the addition of the fluids is important. Outside of the milk cure, some weak stomachs can take milk diluted with water, and assimilate it, where straight milk disagrees. Infants are usually given milk largely diluted with water, but a healthy infant can generally take pure milk without trouble. In any case the water should be gradually reduced and omitted as soon as possible. Skimmed milk is often easier for a weak stomach to digest than full milk. Milk from which the cream has been extracted by a centrifugal separator is better for dietetic purposes than ordinary skim milk, and either is preferable to milk diluted with water.

The taking of at least one or two drinks during the night is a valuable assistance in getting down the necessary quantity of milk. Constipated people should never omit this. It is not necessary to give the stomach a rest while taking the milk cure. It does not need a rest on an exclusive milk diet, any more than a baby's stomach does. Other organs are resting, but the stomach is being built up to a state of maximum efficiency. It is very unwise to omit drinking the milk at a regular time because you don't feel like taking it. If the stomach has been out of order for a long time, there may be a good many disagreeable symptoms, such as bad taste in the mouth, thick coat on the tongue, gas on the stomach, with considerable pain, nausea, and even vomiting, but none of these should prevent the patient taking the regular drinks. The omission of a glass or two, instead of making one feel more comfortable, really has the opposite effect, because the constant, regular procession of milk through the alimentary canal is interfered with, and it begins to "come back," when, if the milk was kept going down, it would carry the gas down with it into the intestines, where it belongs. If, in a case of this kind, the milk is stopped for some hours, all disagreeable symptoms cease, and the patient will find he has a better stomach than he had before starting the treatment, but the cure has only been a partial one, and it may be even harder to get over the critical point next time.

It takes a long experience in this work to give one the necessary confidence to tell a patient to continue the treatment under these circumstances, but it is an absolute fact that I have never seen any harm result from sticking to the diet, (while resting,) and the troubles are only the natural explosions due to the revolution going on in the stomach. If there is any better way to cure an old chronic case of indigestion, with a shrivelled up, weakened and almost juiceless stomach, I never discovered it.

If the patient is lacking in will power, and cannot, or will not, take the regular amount of milk each time, it is a great deal better to take half a glass than none, and resume the full amount at the earliest possible opportunity.

Fortunately, there are very few people who have such a hard time on the milk diet, and they are most all elderly people who have been in ill-health for many years. But even in this class of cases, less than two per cent. have failed to carry on the treatment to a satisfactory result.

The patient must have a warm bath daily, and it is usually taken in the forenoon. Where there is any tendency to insomnia the bath can be given in the evening and usually has a good effect in overcoming that trouble. The first bath should last only 15 or 20 minutes, increasing the time about 15 minutes every day until the patient is staying in the water at least one hour. Use no soap in the bath.

The bath should be prepared with a temperature of 94 or 95 F., and, as soon as the bather becomes accustomed to this sudden change from the air temperature, he should gradually add hot water until he feels perfectly comfortable, neither hot nor cold. The thermometer will then indicate about 98 or 99, although people differ several degrees in their sensations. The temperature must be kept at this point until nearly ready to leave the bath; then enough hot water should be added to produce a thoroughly warm feeling throughout the body.

These three items in regard to the water temperature must be remembered:

Start slightly below body temperature,

Increase to the body temperature,

Finish almost hot, (but never enough to cause dizziness.)

In very hot weather I have found it wise to reduce the temperature of the bath at the start a few degrees—to abstract heat instead of adding it. The principle to be followed is to keep the patient entirely comfortable, and if, for any reason, he is not comfortable, he should get out of the bath.

Having the water too hot on entering will cause a slight attack of indigestion, in the same way that a hot bath affects one when taken too soon after a full meal.

The bathroom must be ventilated in every way possible, and the milk taken at the regular time while bathing.

Ladies who object to wetting the hair can wear rubber bathing caps, but it is better to do without them. The circulation of blood in the scalp is so much greater than usual that the warm skin dries the head rapidly and there is no discomfort where the hair does not have to be "put up" immediately. With very serious cases it is better to cut the hair to a convenient length; it grows rapidly and will be much stronger. In any case the hair ceases to fall out, for it responds quickly to the general condition of the body.

Regarding bathing by females during the menstrual period, I will say that I have never known of any harmful result from the practice, but if they prefer, the baths may be omitted for a few days at this time.

One seldom need be afraid of putting the ears under water. If the eardrum is perforated the ear can be plugged temporarily with cotton. I have seen deafness unexpectedly cured by the combined diet and bathing while undergoing treatment for other diseases.

The proper way to take the bath is to have enough water to submerge all the body except the face and lie at perfect ease with all muscles relaxed and the shoulders supported by the sloping head of the tub, or some contrivance such as a water bag, air cushion or canvas strap.

Breathe deeply and occasionally sink the face under water, closing the nostrils, if necessary, with the thumb and finger.

On finishing the bath do not use cold water or the shower bath, and if possible avoid draughts of cold air, not from any danger of "catching cold," but to prevent the stimulation to the skin.

The reasons for giving baths of this description, and the benefits derived from them, will be explained in another place. I will only say here that I would not undertake to give the milk and rest cure without the aid of these baths.

On getting out of the tub, the patient should dry himself with a soft towel, without unnecessary rubbing, or exercise, put on his bathrobe, and return at once to bed. Weak patients may have the aid of a nurse in drying the skin and returning to their apartments.

I think the minimum time for a milk diet course should be four weeks. Three weeks should be devoted to the rest cure, and the remaining week will be sufficient to gradually get the patient up, and on solid food. In a considerable number of cases patients may continue using milk as a diet, if their circumstances permit, after resuming their occupations, or ordinary habits. I have letters from different people who state that they have lived on milk for long periods, often several years, in one case 21 years. All these persons began the use of milk for some serious ailment, and yet every one of them seems to be in a state of vigorous health and vitality now. The case of Dr. Herman Schwartz, an Austrian physician, who has lived on milk exclusively for 23 years, is interesting, as from all accounts he is in the best of health and strength. He is said to take three gallons daily. One of the best public speakers on the Pacific Coast has lived wholly on milk for four years.

If milk can be taken often enough one can endure more cold than on any other diet. I have lived in the open air in winter with patients where we had to thaw the milk before we could use it. You can get more energy and heat out of a quart of milk than an Esquimo can out of a pound of blubber.

I can state here as a positive fact that an immense amount of physical or mental labor may be done on a milk diet. A young friend of mine lived on about five quarts of milk per day during two terms of college just before graduation and won second honors in a class of over 300, and finished in fine physical condition. His board cost him about $10 per month.

Prof. Weir Mitchell in "Fat and Blood," page 125, says: "I have seen several times active men, even laboring men, live for long periods on milk, with no loss of weight; but large quantities have to be used—two and a half to three gallons daily. A gentleman, a diabetic, was under my observation for fifteen years, during the whole of which time he took no other food but milk and carried on a large and prosperous business. Milk may, therefore, be safely asserted to be a sufficient food in itself, even for an adult, if only enough of it be taken."

The gifted writer, Mrs. Ella Wheeler Wilcox, says: "I believe in the milk diet, because I have taken it with results so marvelous, and so beneficial, that all Mr. Rockefeller's money could not repay me, were I deprived of the knowledge that I gained by the experience. A man of my acquaintance who destroyed his digestion by years of wrong habits, has lived for the last five years in perfect health and strength on milk alone. He is able to work more hours with less fatigue, than any of his acquaintances. He possesses a marvelous complexion and is never ill.

"Another friend who had been a hopeless invalid for ten years, through complications of diseases, has lived on milk for three years, and finds herself perfectly well unless she attempts to return to solid foods. A dozen skilled physicians failed to give her even three days of health, until she gave up foods for milk. Seventeen other personal friends restored their health, and the ability to digest a natural, varied diet, by taking the milk treatment for a few weeks."

Where it is the intention of patients to keep on with the milk diet for very long after stopping the rest cure, it is advisable for them to take larger doses, at less frequent intervals.

Some patients who had been a long time on a half-hour schedule, with a corresponding amount of milk, have said that they found it difficult to eat enough food at one time to last them until the next meal. There is usually no difficulty in taking twice the regular quantity of milk every hour instead of half as much every half hour. This gives more time between drinks for exercise, or business affairs, and, I think it tends to fit the stomach better for the distention of a regular meal, when ordinary diet is resumed.

Business men often carry a handbag full of Mason pint fruit jars, containing milk, and drink one of these at convenient times. Quite a number of old patients have been able to drink a quart of milk at one sitting.

But these experiments must not be tried until the stomach is taking the small and frequent doses without any discomfort, and the bowels are moving regularly. If the stomach is not handling easily the smaller drinks, it will be of no use to attempt the larger ones.

A course of four weeks should ordinarily be sufficient to cure any of the following diseases:

Nervous prostration, general debility, mild skin troubles such as pimples, sallowness, wrinkles, etc., simple anemia, catarrh, biliousness, constipation, dyspepsia, indigestion, hay fever, piles, insomnia, ulcer of the stomach, malaria, arterio sclerosis (hard arteries) neuralgia, neurasthenia, tobacco, morphine and cocaine habits and the first stage of diseases like consumption, rheumatism and kidney disease.

In more advanced cases of the last four named or other chronic organic diseases the diet can be continued as long as visible improvement is made, or until cured.

It may be well to say here that there are crises that come on in the course of the treatment, due to the revolution that is taking place in the body. None of them is an indication to stop the milk, quite the contrary. The most common is an eruption on the skin of the face, body or limbs, usually coming out during the second or third week. I have seen large pimples and boils, but none that ever left a scar.

I wish to speak particularly of crises occuring in special organs that are or have been the seat of disease. You may think there is a recurrence of the disease, but do not have the slightest fear. After the inflammation or excitement has subsided the part is always in a better condition and probably entirely healed for the first time.

CHAPTER III.

REACTIONS DURING TREATMENT.

Before taking up the consideration of the different diseases, I will here describe some of the reactions that take place on a full milk diet.

To begin with the circulation of the blood, we notice directly, in every case, a most remarkable change. Within two hours after commencing the diet, the action of the heart will be accelerated, and within 12 to 24 hours there will be a gain of about six beats to the minute. Within two or three days there will be an increase of about twelve beats to the minute; the pulse will be full and bounding; the skin flushed and moist; the capillary circulation under the fingernails, or wherever it may be examined, quick and active. All this takes place with the patient lying as quietly as possible, making no movement unless necessary—conditions under which normally on an ordinary diet, the circulation would be much slower than usual.

No one can deny the benefit of this condition in chronic disease. It is a result sought by every intelligent doctor, knowing that through the circulation only, can chronic disease be cured. None of the usual methods of heart stimulation, such as alcohol or other drugs, exercise, massage, hot and cold baths, inhalations of oxygen, solutions injected into the veins, or transfusion of blood, can equal the results of the milk diet treatment in effect, in permanency, in total lack of danger. This natural, physiological increase of circulation results from the increased amount of blood, created in the natural way, by the stomach and intestines, acting on an easily assimilated food.

There is no dangerous strain on the heart, because the heart itself is the first organ to share in the benefits derived from the better blood circulating through it. Many patients with serious disease of the heart, organic or functional, valvular or nervous disorders, have taken the milk diet and I have never heard of any but good results. There is no reason to expect otherwise with patients having complete rest.

Nor is there any danger to the kidneys, in spite of their greatly increased work, for invalids with badly diseased kidneys take the milk diet successfully. Some patients, it is true, have slight pains in the kidneys during the first days of their treatment. It is always temporary, and due, I think, to a rapid growth of the organs, so rapid that the sensitive covering of the kidneys is stretched tightly at first.

The amount of urine is very much increased by this diet, and no matter what its previous condition, whether highly acid, or loaded with solids or salts in solution, it becomes bland, non-irritating, and almost as clear as water.

The frequency of urination is a little troublesome at first but in a few days the bladder seems to be able to retain a larger quantity without discomfort; more fluid leaves the body in the perspiration, which is increased by the improved capillary circulation in the skin, and probably the lungs throw off more moisture. However, even then many patients will find it necessary to get up in the night once or twice. It is not advisable to hold the urine very long, as a portion of the water may be reabsorbed into the system.

It is really wonderful how the various parts of the body accommodate themselves to the great changes which they undergo on the milk diet. It is only possible because the greatly increased blood supply brings with it all the necessary materials to make these changes, and a plentiful supply of nourishment for every cell, of every tissue.

In ill health there is always one or both of two conditions of the blood, viz.:

Insufficient quantity,
Abnormal quality.

Disease is a result of a disturbance of the mechanism of nutrition. There may have been predisposing or exciting causes in the way of bacteria or heredity, bad food, air, or habits, but as the abnormal condition becomes apparent to us, we see the evidence of some disturbance of the processes of nutrition.

There is a continuous battle on between the forces that build up and the forces that pull down; between the cells that do good and those that do harm: Nature is always endeavoring to maintain a normal standard against any agent or condition that may attempt to alter it. And when temporarily or accidentally that standard may be departed from, we see immediately an attempt to repair the damage.

No matter what the abnormal condition may be, whether a cut or bruise of the skin, an ulcer in the lung, or the presence of some poison in the system, there is a continuous effort on the part of the natural forces, always acting through the circulation, to restore the normal condition, and we can assist that effort by supplying a proper quantity of such food as may be easily turned into blood.

On the condition of the blood depends the outcome of the struggle, whether life or death, a short or long illness.

The circulation of the blood is nature's agent in eliminating disease, and, increasing the quantity and rapidity of the blood current while improving its quality will assist that elimination.

In a great many maladies,. whether caused by errors of diet or not, the digestive or bloodmaking power is weakened, and to continue the usual food, or to take mixtures of meat, eggs, starchy materials and various drinks, including milk and alcoholic beverages, increases the burden on organs already overtaxed.

If, in addition to the mixed diet, the patient is given medicines for the relief of pain, or for the reduction of temperature, stimulants or sedatives for the heart, cathartics for the bowels or diuretics for the kidneys, expectorants and emetics, hypnotics and narcotics, etc., etc., any one or more of them, the problem for the circulation to solve becomes indeed a complex one, for each and every medicine must act through the blood, whether given by the stomach or through the skin. Even such a simple hygienic measure as bathing, by bringing the blood to the skin and away from the internal organs, interferes with digestion, if that process is not already completed, or of the most simple character.

The action of the heart, as I have said, is always accelerated, soon after commencing the milk diet. There is no reaction from this condition. The effect continues with the diet, but after a varying time the heart may slow down a little because it has become strong enough to do the work with fewer pulsations. The arteries continue full. The heart hypertrophies physiologically, just as a woman's heart does in her first pregnancy. I have observed it many times.

Every organ in the body is hyperemic, or congested with blood, when in active operation, and as the activity increases, so does the blood supply.

There can be no growth, or rebuilding, or regenerating of any portion of the body, without an amount of blood being present in excess of the ordinary tissue-nourishing quantity.

A condition of anemia, or lack of blood, will never be found, when the body is successfully overcoming disease.

We hear a great deal, in these days, of hyperemia as a curative agent, following the ideas of Prof. Bier, and using hot air apparatus to cause a local congestion of the diseased parts.

The use of such apparatus indicates that the natural circulation is defective and unable to push the necessary amount of blood into the part.

But in thus interfering with the circulation, how can we be sure that we are improving matters?

Do we know how to force just the proper amount of blood to a diseased part?

Where does the blood come from?

Is not the remainder of the body weakened, or left without protection?

Does not such apparatus bring the blood more to the surface and away from the deeper and perhaps diseased parts?

Why not increase the blood supply naturally all over the body? Why use apparatus to cause a local congestion when there is a well-known function of the body to attend to just such things, if given the material to work with?

When we suffer an injury to any portion of the body, such as a bruise, a burn, a foreign body needing removal, or the presence of irritating bacteria, or their products, we do not have to wait for the application of any artificial apparatus. The congestion begins at once, through the vaso-motor system, ordered and controlled by the sensory and sympathetic systems of nerves. There is never any mistake about it; the congestion appears promptly in exactly the right spot and no other.

Suppose harmful material has gained access to the circulation, be it chemical, bacterial, or simply a loading up with the natural poisons of the body which have failed to be eliminated. Fever results. Fever is only a name for general hyperemia, and hyperemia is absolutely necessary to throw off or neutralize the poison.

If there is enough healthy blood present in the circulation, or if it is manufactured as rapidly as may be required to carry off the poisons, the system is able to overcome the danger and restore the normal condition.

New and healthy blood is necessary to perform cures; old blood, stagnant blood, impure blood, (from improper foods) no matter how much of it there may be, is ineffective.

In dropsical effusions there is always plenty of blood fluid, but of such a character that the hyperemia set up to repel disease only makes the tissues waterlogged.

Every case of dropsy that has taken the milk diet in my experience, has received immediate and permanent relief.

The heart beats vary greatly in number

in different persons. I have started several patients on the milk diet whose customary pulse rate was around 40 per minute. One lady started with 36, and before the end of the first week showed about 75 per minute, while resting in bed and exerting herself as little as possible. From being a chronic invalid, almost bedridden, weak, listless, almost bloodless, without appetite, she became a strong well woman, and has never lapsed to her former condition.

In patients with fever and rapid pulse there is usually a slowing of the heart and nearly always a reduction of the temperature. This effect is chiefly caused by the larger blood current more easily removing the fever products, and by the cooling of the blood through dilatation of the cutaneous blood vessels, and by increase of perspiration.

It is very unusual for a patient to have a temperature above the normal while on milk and resting, no matter what the previous condition may have been. If the fever does not fall below 100 soon after the patient's bowels are moving naturally, a serious condition is indicated.

The stimulation of a full milk diet is very

similar to the primary effects of alcoholic stimulation on the circulation, but the after results are entirely different, due to the fact that the blood carries with it the food necessary to repair the increased tissue waste.

Stimulation by alcohol is followed by a period of depression which is impossible with milk. Continuous stimulation by alcohol causes inco-ordination of muscles which never follows that of milk. Indeed, the spasmodic, uncertain movements of the hand in writer's cramp may be permanently cured by a proper milk diet.

The effect on the lungs is to quicken the breathing at first; then as the respiratory muscles strengthen, the inhalations become deeper. No matter what disease one may have, the breathing capacity is increased. The circumference of the chest enlarges and the measurement on inspiration increases week by week over that of expiration. The enlargement is too great to be accounted for by increase of muscular tissue or subcutaneous fat around the chest.

These changes, remember, take place while the patient is taking the rest cure. The muscles all over the body increase in size.

To one who has had no experience with this treatment, it seems incredible that the muscles should not only rapidly increase in size, but become much harder. Yet it is a positive fact that the voluntary muscles of the body become firm and solid, almost like an athlete's limbs after a hard course of training. And this, too, while the patient is lying abed all the time, except when attending to the necessary calls of nature, or taking the daily bath. And that bath a warm one, usually considered weakening!

People are too apt to compare a patient taking the rest cure with one in the last stages of chronic disease, or bedridden from the weakness accompanying typhoid or other fevers.

As a matter of fact the two conditions are entirely different. In the latter case, the patient is compelled to take to his bed because he is ill and weak and unable to take or assimilate nourishment, and the food that is given him does little or no good, and may be really harmful, as he has no appetite and lacks the necessary secretions to properly digest food.

But the great majority of patients taking the milk cure are "walking cases." Indeed, many of them demur at the idea of going to bed at first. But go to bed they must to take the milk properly, and, after the preliminary fast, they usually have all the necessary appetite and the right condition of the stomach, to take milk easily, and taking the amount usually given, they are taking more nourishment than the ordinary person takes, even while doing hard work. But milk is, so far as I know, the only food that can be taken in full amount with benefit, while enjoying as perfect rest as may be possible.

The hardness of the muscles, on a milk diet, is due largely to the fact that they are pumped full of blood, like all the other organs of the body.

And, it is well to recall the fact that the internal organs themselves contain a great system of muscles. Not voluntary muscles, it is true, but muscles that are controlled by the wonderful sympathetic system of nerves; muscles which do their work without any effort or knowledge of the will; muscles that work while we sleep. Without them digestion would be impossible, for every movement of the stomach and intestines and the food products contained in them is due to these muscles.

The whole alimentary canal, (œsophagus, stomach and intestines,) contains in its wall a double layer of these involuntary muscle fibers, part of them circling around the organ, and the remainder distributed lengthwise, and through the combined efforts of the fibers, the contents of the canal are mixed with the various digestive juices, and gradually pushed onward until absorption has taken place, and the residue has been expelled.

That is what takes place in the healthy individual without any consciousness or assistance on his part. But in the invalid these small but important muscles are thin and weak and unable to do their duty. It is not unreasonable to suppose that they build up and resume their normal functions just as do the external muscles, while a person is on a full milk diet. Indeed, although we cannot see these internal muscles develop, we have plenty of evidence that they do, as for instance, the increased size of the abdomen, the larger capacity for food, the facility with which the unusual amount of fluid and solid matter is handled, and the much larger amount of feces discharged.

This increased power of the intestinal muscles, and the restoration of the peristaltic movement, is all that is necessary, in many cases, to overcome constipation.

All the muscles, as I stated above, increase in size. I have noted an increase in the thigh of over an inch in a week. The abdomen is always first to show an enlargement, then the thighs and buttocks, although at the same time the neck, shoulders, arms and face are making visible progress. The calves do not make a corresponding gain while the patient is resting, but rapidly assume proper proportions on being called on to support the body in walking.

The rapid increase in girth of the abdomen is very significant. It means that the thirty-odd feet of the alimentary canal are being developed. From the condition so often seen at autopsies where persons have died of malnutrition and the intestines are thin and juiceless, perhaps as brown and almost as dry as the casing from a bologna sausage, they are changing to the thick, juicy and normal condition of an infant's bowels.

The circulation of capillary blood vessels and lymphatics in and around the intestines is greatly increased on a milk diet. The fat of milk is in such minute globules that it is ready for absorption by the lacteals and from them it is carried almost directly into the venous circulation. The large amount of fluid in milk which must pass through the blood before leaving the body, the greatly increased amount of fat, sugar, nitrogenous matters and salts in the right proportion and condition required for nourishment, stimulate the millions of glands lining the canal and they are compelled to increase in size and capacity.

This abdominal increase is very largely in the walls of the stomach and intestines at first. Later on there will be more or less fat deposited subcutaneously. Every healthy person has a protecting pad of fatty tissue in front of the intestines and stomach.

This intestinal development and enlargement is necessary to insure proper digestion and assimilation, but is occasionally objected to by ladies who note the loss of a wasplike waist and the necessity for a new wardrobe with regret. They do not, however, object to the increase in the size and symmetry of the limbs and bust, the filling in of the hollows in the neck, the smoothing out of the facial wrinkles and the "peaches and cream" complexion that goes with it.

I am able to offer some comfort by informing them that a portion of the waist development will disappear when they become more active and another portion will be lost when they quit the milk diet, but with a correct manner of living and sufficient nutritious food the stomach will never return to the previous abnormal condition.

During the last few years, I have noticed an increased number of invalids with disorders of the bowels, principally the colon. These people have usually given a history of "catarrh," or "inflamation of the bowels," often of constipation, and rarely of diarrhea. Some of them say they are full of acid and rheumatic. In a small percentage of these cases, I notice the amount of milk they can take is limited, because an excess brings on diarrhea. They may take three and one half quarts of full milk daily, but another pint causes a watery, acid, diarrhea.

In the case of a patient commencing the

milk diet and taking about six quarts, should diarrhea occur and continue more than thirty-six hours, with passages loose, sour, or greenish, or containing small undigested curds, it is evident that the bowels are unable to digest all the food. The amount, therefore, must be reduced about half until solid movements occur only once or twice daily. Too little will cause constipation, and then the amount must be increased. I have seen cases where a variation in the daily amount taken, of two glasses, would make the difference between constipation and diarrhea.

When the proper dose is found, these patients receive great and lasting benefits. On less than four quarts, the gain in weight is about half that of the usual milk diet, and this is rather discouraging to them, but in every case of this kind where the quantity of milk has been carefully adjusted to the condition of the bowels, the ultimate result has been very satisfactory.

Several ladies who were below weight and affected in this way only gained about two pounds a week while on milk, but on returning to ordinary habits and diet, continued to gain at an even greater rate, and remained free from troubles for which they took the milk cure.

The skin, including the hair and nails, shows decided reactions in the milk cure. A healthy skin is a rarity nowadays, and the average candidate for the milk cure, with bad digestion, poor circulation; and probably kidney trouble, shows plain evidence of his internal disease by a great variety of skin disorders, ranging from the pale, white, and the dry leathery skin, to various forms of eruptions and inflammatory conditions.

Remarkable changes take place in this important organ. The capillary circulation grows faster, perhaps, in the skin, than in any other part of the body. The prolonged warm baths greatly assist in this improvement, by softening up the dead cells of the external layers, and by the moisture and warmth penetrating to the deeper layers. No matter how cold or dry or flabby or wrinkled the skin may be, between the warm baths externally and the increased amount of blood internally, the skin always seems to get back to a healthy condition. Patients who had not visibly perspired for years, show a perceptible sweat within a few days, and frequently the skin starts up action suddenly a short time after the patient has gone to sleep in the evening and he wakes up bathed in perspiration. I have seen such cases where not only the bed linen, but the mattress as well, were so soaked with sweat as to require changing.

Such a climax, weakening and discouraging to some invalids on an ordinary diet, has just the opposite effect on the milk diet, because it is the result of increased capillary circulation, and not due to weakness of the blood vessels and thin, watery blood, as in the ordinary "night-sweats."

If the warm, moist skin be rubbed, soon after starting on the milk diet, one can often notice little black rolls of dirt, dead cells, and waste matter discharged through the sweat glands, and the odor coming from the skin saturates the atmosphere of the room, and will be found excessively strong on opening the bed to air, especially with rheumatic patients. Indeed, the rooms of these patients smell like a vinegar factory for a few days.

A rapid increase in body weight occurs to every one taking the milk diet, no matter what their previous condition or disease. While this is usually welcome, there are certain patients who do not desire it, but they have to accept it, at first anyway, because it is impossible to take the cure correctly without gaining in weight.

Someone has divided the human race into two classes—those who are too fat, and those who are too thin, and while the milk cure appeals more to the latter class, yet it seems to me that stout people get just as much benefit from it as thin ones, but it is harder to induce them to take it. The gain in weight made by a person who is overweight or about normal, is not as great as that made by a thin or emaciated person. The latter will take on weight rapidly, almost as a sponge soaks up water. Most of them are poorly nourished, whether they are eating much or little, and the milk alone, taken under proper conditions, seems to be just what they need, and they build up all parts of the body very easily. The average gain in weight is about five pounds the first week, and after that about half a pound daily. This latter rate continues for weeks, or months, until they are near the normal weight. The greater increase for the first week is, in some measure, due to the fact that they have had more or less of a fast before commencing the diet, and are consequently almost empty, and in a good condition to assimilate nearly all of the milk.

A gain of twelve pounds the first seven days has been made under my observation, and only recently a young man gained ten and a quarter pounds in his first three days, but his was a very exceptional case, as his stomach had been in such a wretched state that he had been unable to retain even the simplest food previous to taking the treatment. I started by giving him small and frequent doses of carefully warmed milk, while he was resting completely, in bed. Beyond a very slight pain in the stomach at first, he had no discomfort at any time, and rapidly regained his health.

I am often asked if the average rapid gain of flesh is not too great to form healthy tissues, and even if it may not be unsafe.

I say, emphatically, that all this increased weight is made up of healthy tissues and that there is absolutely no danger while taking complete rest.

There are certain preparations advertised, by the use of which, it is claimed, rapid gain is made in weight, while eating ordinary foods. I have seen very injurious effects from the use of some of these drugs, and I regard such methods as wholly unnatural. The flesh gained is probably largely fat, and the digestive organs, instead of being built up and fitted for normal digestion, are worse off than before taking the medicine.

The gain made on milk diet, while resting, is not principally fat, as some people imagine. An increase of an inch a week in an emaciated person's thigh, between the knee and the hip, cannot be called fat. There is very little fat in this part at any time, but there is an enormous group of muscles and it is the growth of these muscles that produces the enlargement. The muscles increase because they are distended with blood.

The inunction of fat, or so-called flesh foods, or "oil rubs," cannot produce any permanent benefit and may cause considerable harm. The massage may temporarily stimulate the circulation, but it would be better to practice it without the oil.

I recall only a few cases occurring in my experience which lost weight in the first week, while taking the amount of milk I had prescribed.

Two of these suffered from valvular disease of the heart, and after the initial loss went on as usual, gaining weight and health.

Another case had been the subject of severe surgical operations and lost several pounds at first, but then gained at a fairly satisfactory rate.

Two gentlemen suffering from diabetes lost weight on the milk diet (6 quarts) for a few days and both quit the treatment, but seemed to have derived some benefit from the short course.

CHAPTER IV.

DYSPEPSIA.

Probably the simplest trouble that may be treated by the milk cure is dyspepsia, or indigestion.

Usually brought on by an incorrect way of living, when the condition is relieved and the patient instructed how to avoid a return of the disorder, there is no good reason why he should ever suffer from it again.

When you tell people that they are not eating right, they may say they eat the same as other members of the family who are apparently having no trouble. Now, it is a fact that there are no two people with stomachs just alike. Among thousands of cases who have taken the milk diet under my observation, I do not remember any two who acted, or reacted just the same. With a score of patients taking the milk diet at the same time, with all conditions as to amount of milk, time and manner of taking just the same; all resting in bed, all bathing daily, everything just the same, as nearly as may be, yet no two will have the same symptoms.

With some the bowels are constipated; others have a diarrhea, some have regular movements. Most people have a chalky white color to the stools at first; some start off with a normal yellow color. Some of those with white stools will suddenly change to a deeper tint, as the liver starts up; others make the change very gradually.

Some patients complain of a bad taste in the mouth; others never have it. Most of them have a heavy coating on the tongue, at first brownish or yellow, later white; others manage to keep their tongues clean. Great variety is observed as to the way the milk tastes. Most patients seem to have a relish for it, in greater or less degree, others declare it is just like taking medicine and they only take it for the effect. Some will say it tastes very sweet, like sugar; others that it has a bitter or sour taste. Precisely the same milk, mind you. The disagreeable taste may disappear at any time, and perhaps not return.

It is very seldom that patients become hungry on the milk diet, and where they do, it probably is because they are not taking their milk regularly. But there have been cases who felt like eating most of the time during the first two weeks. In these rare cases the craving has been generally for some special article of diet, as bread, or some kind of vegetable or fruit. I have never known anyone to desire meat, except once or twice a wish was expressed for fat, crisp bacon.

Thirst is a very rare symptom while on the milk diet, and I do not remember any patients in New York who took water in addition to their milk, but in California I have seen several such cases, perhaps due to the drier atmosphere. There is probably no reason why water should not to taken during the treatment, but I hardly see the need of it, as most patients get over five quarts of water in their milk daily.

It is not uncommon for those taking the milk to wish for something sour, and particularly sour fruit, apples, oranges and even lemons. Others prefer the sweet fruits, peaches, plums, grapes and melons. Nearly everyone relishes dried fruits like figs, dates, prunes and apricots, and all these are frequently useful in overcoming the initial constipation.

In the matter of sleep patients vary widely; some want to sleep all the time, while others only sleep a few hours at night. Cases of insomnia sometimes notice no improvement for several nights, and then, all at once, they begin to sleep like children.

I have taken considerable space to explain how the milk diet affects different people, while speaking of dyspepsia, because nearly every disease is accompanied by more or less stomach trouble, although the symptoms are quite varied.

Indigestion is almost a national disease with Americans, and in very many cases it is due to imperfect breathing, or lack of exercise, and overeating.

There is not much use of my wasting the reader's time in giving advice as to the amount of food they should eat, or what kinds they must avoid and what they may eat, and how long they should chew it, and how many meals they should take, because all dyspeptics have had plenty of such advice, without being cured, but I will, later on, give some directions to be followed after putting the stomach in good order by the milk diet.

I firmly believe that defective breathing is more of a cause of dyspepsia than over-feeding. Few people realize how important breathing is to health. We breathe mainly to absorb oxygen. The function of oxygen is to combine with the food we eat, and if sufficient oxygen is not taken into the system to oxidize the food, indigestion results.

Food is taken into the body just as fuel is taken into a furnace, for the same purpose—to be burned up, and burning always means oxidation. A lamp or a stove cannot burn without a plentiful supply of oxygen, nor can the human body perform its functions more than a minute or two without air.

Oxygen should really be considered a food, for none of the regular foods would be of any use in the body unless they combined with oxygen.

Some people never breathe right; many people work and sleep in places where the air is bad, and, while it is possible for either class to enjoy fairly good health, if the defective breathers have to breathe the bad air, the result is always ill health.

A man may work every day in a place where the air is impure and lacking in oxygen, and yet, if his work calls for vigorous exercise, and therefore copious breathing, he may appear to be in the best of health.

But let the shallow breather work in the same place, at some sedentary occupation, and before long his health fails, he becomes pale-faced, anemic, has less strength, less endurance. His desire for food decreases, and what he does take is not thoroughly digested, hence he has less blood, and that of a poorer quality. Perhaps, realizing that his stomach is not performing its functions properly, he assists it with some digestive medicine, or he takes foods that are recommended to him because they are predigested. In either case he may notice an apparent improvement, but in either case he has further weakened his stomach by usurping its natural functions, and if the primary cause of the trouble is not remedied, his temporary expedients will soon fail to have even an apparent or transient effect.

I want to make this point clear: **We cannot habitually perform for the body any of its functions that should naturally be performed** unaided, without weakening the part concerned.

We cannot use massage or kneading of the bowels for any length of time to produce defecation, without weakening the natural peristaltic movement of the intestines; nor can we use cathartic medicines long without the same result.

The stomach should digest our food, and we cannot live on predigested foods long, without weakening those glands which normally secrete the digestive juices.

We certainly cannot add pepsin to our food before we eat it, without taking away the function of the peptic glands, and, while they may have been secreting too little pepsin before, they are likely to produce still less when the food comes into the stomach already peptonized.

The simple act of cooking, which is one kind of predigesting, may, in some cases, be a contributary cause of weak digestion. Our digestive apparatus was originally designed to work on uncooked foods, for fire was a later invention, and all animals at the present time, except man, use by preference uncooked foods.

We are too much inclined, in the hurry and

worry of modern life, to eat those things that may be swallowed and digested quickly, without regard to the ultimate effect on the stomach.

Prof. Einhorn says: "The diet in health should not always comprise the most easily digestible substances. For by so doing we weaken our digestive system."

Stomachs can be spoiled by giving them too little to do, and they must be able to digest much that is difficult of digestion, as well as that which is easily digested.

But, aside from the fact that the stomach must be able to take care of such foods as come to it on an ordinary diet, the great question remains, are these predigested foods able to make as good blood as natural foods do?

It is a fact that the great majority of patients applying for relief from digestive troubles are in the habit of using foods designed to save the stomach some of its natural work.

A story I hear very often runs about this way: "Some time ago my health began to fail; my stomach was bad, and the only thing I could eat with comfort was Somebody's breakfast food (or Dr. So-and-so's prepared food, or some grain combination claiming to be predigested and already to eat.) But while this food caused no distress, I have been getting weaker and weaker, and cannot touch the ordinary food that my family eat."

And these people always lay stress on the statement that they have been very careful of their stomachs!

Most of these prepared cereal foods are steamed or boiled into a mush, with various ingredients like salt, glucose, molasses, or malt, added. Then they are usually either made into a dough and baked, and ground up into crumbs, or rolled into flakes and parched. After being sealed up more or less tightly in pasteboard boxes, they are ready for sale. When finally the retail dealer gets such foods, they may lie on the grocery shelves for months before being sold. Every country store is stacked to the ceiling with preparations of this kind, for which an artificial demand was created by enormous advertising, but when the advertising stops, so does the demand.

The manufacturers of many of these products, who have become rich by buying cheap cereals, or grains that have already been used in making malt liquors, and selling them for ten times their cost, employ high-salaried advertisement writers, who dilate on the cleanliness and thoroughness with which the goods are handled and cooked, but I do not believe that stuff prepared in this manner can be of much service in the human stomach, and even animals refuse it unless they are very hungry.

I do not include in this class foods like Germea, which is not cooked, nor sterilized by any chemical method, nor the rolled preparations of wheat, oats, and rye. The latter are steamed for some time, and while still wet are run between rollers and pressed into thin flakes. After drying the product is ready for marketing. Such rolled grains do not pretend to be more than partly cooked and are supposed to be thoroughly recooked before serving. Grain prepared in this manner does not lose its vitality, or blood-making power. Some grocers raise the objection that such foods do not keep for long periods, like the ready-to-eat, sterilized brands, as they are apt to be attacked by weevils and other insects. This, in my opinion, is a pretty good test of the food quality of an article. Insects, with their magnified sense of sight, smell, and taste, are better judges of the food value of an article than human beings. As an example, I have known ants to find an opened package of Germea on a pantry shelf, and when discovered, the wise little animals had a line many yards long between the cereal and their nest. Each ant returning to the nest carried a little particle of the food, doubtless for the nourishment of the home colony. An instructive feature of the incident was the fact that the ants, to get at the preferred article, had to climb over several opened packages of other foods, each of which was guaranteed by the manufacturer to be all ready to eat.

The cause of much indigestion, particularly of starchy foods, is deficient secretion of pancreatic fluid. Digestion of starchy foods is either performed by the action of the saliva, in the mouth and œsophagus, or after passing through the stomach, by the juice secreted by the pancreas. The stomach itself has no action on starch.

The pancreas is very often at fault in people of sedentary habits, and, if such people do not very thoroughly chew their bread, potatoes, etc., and thereby largely digest the starch before it is swallowed, it passes through the stomach unchanged and is very apt to ferment in the intestine.

Persons with this trouble can often live in comfort on meat, eggs, fruits, and non-starchy vegetables, but bread, for them, is by no means the staff of life.

The secretion of a healthy pancreas, which is discharged into the small intestine, amounts to about a pint in 24 hours. This amount may be greatly reduced, in ill health, through inactivity of the gland. Total absence of the juice, from disease or removal of the gland, results fatally.

The pancreatic juice is of great importance in the digestion of milk, and on a milk diet, the gland becomes very active, and presumably returns to a healthy condition, for, on resuming a normal diet, patients do not suffer from starchy indigestion and fermentation.

Other forms of indigestion, or inability to digest certain foods, as berries or acid fruits, nuts, and certain vegetables like cabbage, are always completely cured by this treatment.

To sum the matter up in a few words, it puts the stomach in a normal, healthy condition.

CHAPTER V.

CONSTIPATION.

Constipation is a very frequent accompaniment of digestive disturbances, and while it is usually only a symptom in itself, and disappears as the trouble which caused it is cured, it deserves separate consideration.

It is my impression that a great deal of constipation, or irregularity of the stools, is due to the fact that many people do not know how to attend to the important function of defecation. They either do not understand it, or they wilfully neglect it.

A little study of the parts involved, and their physiological action will be interesting and instructive.

Food, after passing through the small intestine rather rapidly, enters the colon, or large intestine, as a liquid, or of a liquid consistency. The fluid is largely absorbed during the slower passage through the large intestine, leaving a residue of feces to be discharged from the rectum.

The large intestine is about five feet long, including the rectum; which comprises the last eight inches. The rectum, like the rest of the intestinal canal, has involuntary muscles in its walls, running both longitudinally and circularly. The circular musclar fibers near the outlet are increased in thickness and form a well defined ring, about one inch wide, called the internal sphincter muscle. Just beyond this, but entirely separate from it, is the external sphincter, of voluntary fibers, which ordinarily keeps the anus closed.

Now notice the different kind of muscles composing the internal and external sphincters. The external sphincter is a sort of purse-string muscle, under control of the will, which keeps the outlet closed except when we wish to discharge the contents of the rectum. The internal sphincter, is, like all the rest of the muscular fibers in the intestinal wall, an involuntary muscle, and we cannot directly compel it to open or shut by will. power, no matter how much we may desire it.

The contents of the intestinal canal are propelled onward by peristaltic, or worm-like movements, which are entirely involuntary.

These movements are caused by wave-like

contractions of the muscles in the walls of the tubes, each part of the tube as the wave reaches it, narrowing its caliber, and then gradually relaxing and dilating. This wave of contraction is gentle, and progresses slowly from above downword. The advancing wave is always preceded by a wave of relaxation, or inhibition. When we inhibit or relax the tension of the muscular fibers in any circular organ, as intestine or bloodvessel, the organ naturally dilates and the space in its center becomes larger.

As the contents of the large intestine arrive at the rectum they are composed of undigested and indigestible matter, about 75 per cent. of water, and considerable waste mater, including cast-off cells, inorganic salts, putrid products, and bacteria. When a sufficient quantity of feces has arrived in the rectum there is felt a need of expelling it. This sensation varies greatly, according to the amount of matter present, and the susceptibility of the individual, but principally owing to the nature of the discharge. If the matter is rather solid, and of a non-irritating nature, it may be retained in the rectum for hours, or even days, while if it be watery or acrid, as in diarrhea, the strongest effort of the will is sometimes insufficient to keep the external sphincter closed and prevent a passage of the contents.

The act of defecation is normally an involuntary one, as may be seen in infants and animals on a natural diet. In most adults it becomes partly voluntary, owing to a variety of causes, such as the habit of preventing the stool until a convenient time, and to unhygienic habits in general. The voluntary part is, however, smaller than generally supposed, consisting mainly in the relaxation of the outlet, and the compression of the abdominal contents by holding the breath and contracting the diaphragm and abdominal muscles.

The compression of the abdomen, or straining, practiced by many persons, and particularly constipated people, usually does more harm than good. It has no effect on matter in the lower part of the rectum, as Professor Foster says in his Textbook of Physiology: "A body introduced per anum into the empty rectum is not affected by even forcible contractions of the abdominal walls."

If the peristaltic movement is not operating in the rectum, and the internal sphincter is not properly relaxed, we should not try to have a movement. If, under these unfavorable circumstances, after long straining, we do succeed in our object, what happens is about as follows: We have so compressed the abdominal portion of the large intestine that its contents, perhaps unready for the movement, are squeezed downward, while the absorption of the fluid portion is unnaturally hastened, and matter thrown into the blood that should have been further elaborated in the bowel, or discharged with the feces.

Although a passage may sometimes be effected by such unnatural straining, its results, both immediate and remote, are bad. The immediate effect includes the disarrangement of the digestive processes, not only in the large intestine, but in the small as well, and probably other abdominal organs, in the forcible pushing into the blood and lymphatic circulations of unsuitable substances, causing headaches and auto-intoxication, while the after consequences are that the constipated habit is more firmly fixed, **the next stool almost certainly being dry and hard,** and the natural mechanism of defecation more weakened and less inclined to perform its duty.

Piles, or hemorrhoids, are probably always caused by straining at stool, as the pressure prevents the venous blood returning to the heart and it accumulates in the mucous membrane of the rectum and distends it until tissue gives way and a blood tumor, or pile, results.

The contents of the bowel act in different ways to produce a normal stool. By irritating the mucous membrane, nervous centers are excited which cause a reflex peristaltic movement of the intestinal muscles, and by the secretion, or production of osmotic conditions, which cause fluid to flow into the cavity of the intestine, until it becomes so watery that it may be discharged.

Various laxative foods act directly, on the arrival of their undigested portions at the rectum, as the coarse fibers of cereal coverings, small seeds of fruit, or indigestible skins of tomatoes, prunes, etc., or they may act immediately through the circulation, as it is not uncommon for people to have to go to stool within a short time after eating the first peach, or pear of the season, or any fruit that is unusually well relished. And some people say that eating any article which causes a copious flow of saliva will bring on a movement. A small amount of laxative fruit usually acts better than too large a quantity. Some doctors advise the eating of just one fig at bedtime, or one apple before breakfast, knowing that eating a larger amount, or perhaps overeating, seems to have frequently the opposite of a laxative effect. Overeating of any food or foods is a prolific source of constipation.

While constipation is an unsanitary habit, to say the least, it is nevertheless a fact that many people who have it magnify the condition and its dangers. There are many such persons, whose one object in life seems to be to have a daily movement, whether there is anything to be moved or not, and they are in misery unless they have it. And, after having it, by force, if necessary, they are not happy, because they immediately begin to plan for the next day. Everything they eat or drink is judged by the test: Is it constipating, or not? They stuff themselves with unsuitable foods, because some one declares them laxative, and they decline really nutritious articles, because they may have a reputation for causing costiveness. Their drinks, also, have to pass the same test, and healthful fluids are rejected, perhaps in favor of some foul smelling and tasting mineral water, because the latter moves their bowels, disregarding the fact that these waters are often artificially impregnated with salts by the manufacturers and may contain many impurities. It is but a step from them to out-and-out cathartic medicines. and then, usually, all chance of restoring the normal movements of the bowels is gone, except by radical action.

A simple and good cure for many of these cases, which I have often successfully applied, where the people had common sense and will power enough to carry it out, is to have them eat, in moderation, everything they relish, chewing it well, and instead of trying to have a movement of the bowels, try hard **not** to have one. Instead of yielding to the first faint hint that the bowels might move, restrain it, until the next day, or two or three days, if necessary. Retain even the gas, if any is inclined to pass, and, my word for it, there will come a time when an impulse will be felt about which there can be no mistake, and a satisfactory evacuation will result, without straining or forcing into the circulation a lot of stuff which does not belong there. Don't try to force out every particle that you think may be in the rectum, but keep some for next time, so to speak, and the next time will also be easy, and the next time after that, and all the other times. This plan will not suit the manufacturers of expensive machines intended to wash all the food out of the colon, by using large hot-water injections. They may claim that you will get ptomaine poisoning, or auto-intoxication, but such things, in my experience, have usually been caused by bad food, imperfectly digested, forced out of the colon into the body against the wishes of the absorptive cells, which could not stand the pressure put on them.

As to the use of injections, I will say here, that occasions may arise when they are temporarily of great use, but as a means of curing constipation, it is irrational to distend the bowel, already weakened and dilated, with large enemas of warm, or hot water, and the most difficult cases to cure are those where the colon and rectum are paralyzed from long use of such measures.

There would be very little constipation

with any one, if the internal sphincter relaxed readily. While it is contracted it acts very much like a valve or gate, opening inwardly, and the more pressure we put on it the tighter it is shut.

Just why it does not open, or relax, when we wish it to, is a difficult question to answer. It cannot be relaxed by simply willing it to as we would relax the muscles of the arm, for it is not a voluntary muscle. Some people, who are always on a nervous tension, put so much extra force on the external sphincter all the time that it gets in a state of continuous or tonic contraction, and communicates its rigidity to the internal muscle sympathetically. Such people may suddenly decide that it is time to have a movement, and they relax the outer muscle, and attempt, by pressure, to overcome the resistance of the inner muscle. But this is not the right way to go at it.

A better way is outlined above; simply wait until the desire is irresistible; then the inner sphincter is sure to be relaxed and the peristaltic movement sufficient to move the contents out without using abdominal pressure.

Another, and perhaps, opposite method of

securing the desired result is to obtain a general relaxation of the body, which will secondarily or sympathetically affect the involuntary muscles. This is explained in another part, under the title of Rest.

A method frequently advised is the use of massage over the abdomen, by manipulation with the hands, or rolling a cannon ball over the location of the colon, or bringing the abdomen forcibly against some object, as a strap between two posts. All these are unnatural procedures, and while possibly useful, temporarily, in some cases, they never tend to produce a cure, but instead, further weaken the intestinal muscles by usurping their functions.

As contrasted with passive exercise, active exercise is far better. Exercise of any part of the body makes deeper breathing necessary, and that means more up and down action of the diaphragm, which in turn produces more movements of the abdominal contents. More oxygen enters the blood, and more blood circulates through the vessels everywhere, stimulating all the muscles. This induces warmth, and perspiration, and when there is external perspiration, there is usually a watery excretion through the mucous membrane of the intestines as well, for the inner lining of the intestines is really outside of the body proper, in the same way that the skin is.

Not only does proper exercise bring into use the external muscles of the abdomen, whose action is readily apparent, but also important muscles lining the abdominal cavity, connecting the backbone and pelvis and thighs, the movements of which must have a considerable influence on the abdominal viscera.

There can be no question about the benefits of exercise to the constipated, **it is a necessity.** Movements particularly useful in these cases will be explained under Exercise. Some directions in regard to diet will be given in the chapter entitled After Treatment.

The reader will note that so far, in treating of constipation, I am speaking of persons in ordinary circumstances, and not those on the milk diet, or resting.

What does the milk diet do for people afflicted with constipation? It is the only perfect and natural cure that I know of.

On no other diet can the bowels be restored to their natural functions, while the patient remains in bed, resting.

On no other possible diet can any one build up the entire muscular system of the body, both voluntary and involuntary, while taking complete rest.

On no other diet is it possible for one to gain healthy flesh rapidly, without exercising, or submitting to massage.

Most people find that drinking fresh milk with other food increases the tendency to constipation. Even when they attempt an exclusive milk diet, using a few pints daily, the trouble is increased. A few people of this class say that milk acts as a laxative to them, and the use of buttermilk, sour milk, and sour milk cheese, tends to prevent constipation in nearly every case.

It is impossible to use anything but fresh milk in the milk cure, because that is the only substance that the stomach can take continuously, for unlimited periods, without tiring of or rejecting. However, it is often useful to give a glass or two of buttermilk, or some cottage cheese (made without cooking) if the bowels do not move naturally while taking the full amount of milk. Sour milk that has coagulated, called "clabber," or "loppered milk," may be beaten up with an egg-beater, and makes a very good substitute for buttermilk.

But the best of all ways of overcoming the initial constipation on a milk diet is to take more milk. In every case there is necessarily a considerable portion of the milk undigested, and the percentage of undigested matter increases as we increase the amount taken. On a generous milk diet regular stools occur largely as a mechanical result at first; the accumulated feces are too great in amount to be retained. This is often noticed where patients are taking an amount of milk somewhat too small to cause daily movements. Increasing the amount one or two pints daily generally has an immediate effect, changing the discharge from dry, hard, round balls to a soft, continuous cylinder, with more frequent movements. It is possible that only a small portion of the additional milk is digested, although an increased rate of gain in weight is always shown in such cases.

While it is true that certain parts of the milk, as the fat and casein, or cheese, are never entirely digested, there are other parts, as albumin, milk sugar, and mineral salts, that are completely assimilated. The cells lining the alimentary canal have a selective action; they take out what is needed, and reject the remainder, and, under the natural conditions surrounding the milk cure, it is always better to provide too much food than too little, in order to be sure of getting enough of the absolutely necessary materials. Where only one kind of food is taken, it is a simple matter for the digestive apparatus to select from it the needed ingredients, and pass on the residue. The dream of theorists that some day we may be able to supply all the needs of the body by means of a daily pill and a swallow of liquid, and not have any undigested residue to bother with, will never come true.

Before starting the milk diet, constipated persons should have at least 36 hours fast from ordinary foods, but any ripe, fresh, or dried fruits (except bananas) may be eaten. If the stomach is too weak to handle raw fruits, they may be cooked, without using sugar. Several glasses of water, either hot or cold, should be taken, but do not distend the stomach with it too much, for water, taken in unusual quantities, is almost as indigestible and uncomfortable as some foods are, to a weak stomach. It is not absolutely necessary to have fruit during the fast, and if patients cannot find fruit that they can eat wih a relish, they had better omit it. But where fruit is eaten, it is frequently the case that, on beginning the milk diet, the bowels move spontaneously the first day. If the bowels do not move the first day, let them go until the next morning, unless positive discomfort is experienced, and, if no indications of a movement on the second day, an enema of warm water (at the body temperature) may be taken. No more water should be taken than is necessary to accomplish the purpose, and, if even a small passage is secured, do nothing more until the next morning, unless the bowels move naturally. Do not strain, or attempt to force a movement.

The movement on the morning of the second day of milk drinking will contain the undigested part of the fruit eaten just before starting the milk, and perhaps the last part of it will be tinged white or yellow by the milk. The movement on the third morning may still contain traces of the fruit, but will be mostly a milk stool.

If any trouble is experienced in passing feces from the rectum, even with the aid of an enema, it will be on this third morning, and never after that. It is only a local trouble, right at the outlet of the rectum, and copious injections of warm water, with perhaps a little soapsuds or castor oil, will overcome it, and it will not recur, if the milk is kept going regularly.

It is best to secure an evacuation every day, if only a small one, and the warm water enema will be all that is necessary to obtain it. If the injection is needed for several days, decrease the amount of water used every day, until only about a teacupful is used. The rectum will be getting more power all the time, and probably the feces will become softer.

After about a week of using the warm water injections without securing spontaneous movements, it is a good plan to use a very small injection of cold water, the colder the better. Cold water acts as a stimulant to the rectal muscles and causes a contraction which brings on the movement. Usually two or three trials with cold water will put the bowels in shape to have natural movements.

The color of the stools, after the bowels

are clear of food previously eaten, is a chalky, or grayish white, in nearly every case. A few persons have soft stools, of a yellowish color, right from the start. As the patient goes on with the diet, more yellow color appears, and the discharge is softer, and movements more frequent—two or three a day. A light orange color usually indicates a good condition of the bowels.

When normal evacuations are established, there is never any return to a constipated condition so long as the same conditions of rest, milk, etc., are continued. When ordinary diet and habits are resumed there is very seldom any difficulty, because the regular habits established while on the milk, and the increased power of the peristaltic muscles, the improved digestion, and increased circulation, all tend to prevent a relapse.

There have been patients with obstinate and long-continued constipation, who, for lack of time, could not stick to the milk diet long enough to entirely overcome their trouble, but who found their movements entirely regular on resuming ordinary diet.

Of all the constipated people who have taken the treatment correctly, for at least four weeks, at least 98 per cent, have been completely and permanently cured.

CHAPTER VI.

CONSUMPTION.

The milk diet has, in my experience, been used more by consumptives than any other class of invalids.

This is partly due to the greater prevalence of tuberculosis and partly to the fact that the immediate gain in weight appeals to them.

All consumptives know that a decrease of weight nearly always means an increase in the disease; conversely, they believe increasing weight indicates a gain in health.

Of far greater value than the mere increase in weight is the improvement in the blood, both in quantity and quality, and the changes in the lungs it brings about.

In addition to the increased capacity of the lungs previously spoken of, there are some characteristic reactions which always occur in pulmonary tuberculosis.

Many people contract consumption and are completely cured simply because their vitality is strong enough to overcome the germs.

Just as soon as a consumptive has sufficient blood to prevent the wasting of the tissues, the loss of weight stops, and on a further increase of the nutritive fluids, the healthy cells are so strengthened as to prevent any extension of the disease. Next follows in natural order the recovery of the affected cells, or the development of new ones, and the elimination or absorption of the germs.

Almost every case of tuberculosis with consolidation of lung tissue, starting in on the milk diet, has a coughing crisis, generally about the end of the first week. The cough and expectoration greatly increase as the consolidated portions of the lung loosen up and air enters the cells which have been filled up with the products of the disease.

The cough is easy and the sputum comes up from the lungs readily, while previously the cough may have been hard and ineffective.

Most of the authorities on tuberculosis include among the first signs of improvement a decrease of the cough and expectoration.

This may be the case in a dry climate on an ordinary diet, because the inflammatory products would be gradually absorbed and largely eliminated through other channels or perhaps even made over and used in the system.

Probably this process also occurs to some extent in this treatment but I have seen almost a quart of sputum come from a patient in twenty-four hours at this time.

Later on, of course, the coughing and sputum decrease.

On physical examination at this stage the air is heard entering the previously solid portions of the lungs and as the air cells become cleared out, the breathing sounds become normal. In cases where cavities have formed from the breaking down of the consolidated area, sounds are heard later on, indicating the hardening of the walls of the cavity and cessation of the extension of the disease.

After another period of time the cavity may decrease in size. I have observed cavities as large as a sugar bowl become cleared out and firm walls formed around them and years after have noticed a sinking in of the chest walls and almost complete obliteration of the cavities with normal lung tissue around them.

Air is more important to a consumptive than to any other invalid. In my opinion it does not matter what kind of air so long as it is pure.

If the condition is serious you must stay in the open air day and night. Keep out of rooms, churches, theatres, cars or any crowded place.

Under ordinary hygienic treatment many cases have been cured at high altitudes, others in the Salton sink below sea level, others on long sea voyages or on sea islands, others in the dry air of Arizona and others in the moist air of Cuba and Florida.

On a milk diet or not, a third stage consumptive who has regained his health in a certain climate should stay there.

The lung tissue built up on a Colorado mountain will not long stand the air of New York City. Many, many cases that come to Arizona and get "cured," return to other States to live, and after again losing health, come back to Arizona.

But the "cure" seldom works the second time, never the third.

I say to you, pick out a place where you are satisfied to live and when you regain your health you can remain there and retain it.

One thing I wish to caution third stage

consumptives about. There is only one possible danger to anyone taking the milk cure and that is the rupture of an artery or aneurism from the increased blood pressure.

I heard of a patient who had an aneurism or blood tumor caused by a wound of the large artery of the thigh and on beginning a milk diet the pressure of the blood ruptured the aneurism causing a fatal hemorrhage before medical aid could be summoned.

The same danger would be present after a major surgical operation, and is well understood in hospitals, where such patients are given very little fluid for some days after the operation.

I do not believe this danger would be present with an aneurism caused by disease of the artery, because the new blood would rapidly strengthen and restore the weakened walls. I would not hesitate to apply the milk diet to such a case, beginning it rather gradually and should confidently expect a complete cure, but such patients must have complete rest for several weeks.

I can imagine a patient with a cavity in the lung crossed by an artery, whose walls are eroded and weak, which might break and cause considerable loss of blood.

Out of several hundred lung cases I have never heard of any serious hemorrhages, but where cavities are suspected to exist, the beginning of the diet should be gradual, commencing with, say, three quarts and increasing one glass daily until the regular amount is reached about the third week.

There may be small hemorrhages from capillary vessels which are extending into the diseased tissues, where the circulation had previously been absent, but they amount to nothing. They soon close and the loss of blood is not noticed where it is being made so rapidly.

Bleeding from the mucous membrane of the nose and throat may also occur where there has been a catarrhal condition, but it is always slight and never harmful.

Those living in high altitudes and subject to hemorrhage should always begin milk gradually

CHAPTER VII.

CATARRH AND ASTHMA

Catarrh is a very common disorder, affecting the mucous membranes in various parts of the body. It varies from a slight, transient, "cold in the head," to chronic and serious conditions of the nasal passages, throat, lungs, stomach, intestines, etc.

At first one may be subject to the attacks only at certain times of the year, following exposure to unusual weather conditions, usually after being over-heated, but afterward it may be present, more or less, all the time.

The milk diet treatment seems to have a direct and invariably beneficial effect on catarrhal conditions of any mucous membrane, the very first result is a strengthening and general building up of the softer tissues of the body, those which are first influenced by a richer blood supply, and the cells of which these mucous membranes are composed, are very quickly influenced.

If no serious complications have set in, there is no case of catarrh or hay fever that cannot be **permanently** cured by a four weeks' course of milk diet.

Some of the happiest cases I have had were asthmatics, and I can state positively that, if the disease has not progressed to the stage where the air cells of the lungs have broken down into emphysematous cavities, a complete cure may be made.

At first, asthma is only a nervous disorder, but after years of straining, and wheezing, and "doping," most asthmatics develop emphysema and bronchitis. As with other nervous troubles, here, too, the milk diet is a specific. But patients must throw away their medicine bottles, inhalers, and smokers ,and depend entirely on the milk, with complete rest, and warm baths. The cure cannot take place as long as applications of cocaine, or other "deadeners" are made to the air passages. The first thing to learn to do, after starting the milk, is to relax the whole body, and lie down flat in bed, and this point usually comes even in some bad cases, within the first two or three days. After that time the recovery goes on without interruption. Do not discontinue the rest part of the treatment too soon, but let it bear some relation to the time the disease has existed. As the patient gets stronger, the lungs clearer, and the breathing regular, increase the air in the room and remove some of the covering from the bed. Harden them off before getting up and beginning exercises. Continue the milk for weeks, perhaps, after getting up, if the case has been serious. It is important, even more so than in other diseases, to do the cure completely at the first trial, and not stop, after a certain amount of improvement is obtained, with the expectation of finishing up some other time.

Asthma is easily curable previous to the time that actual breaking down of lung tissue takes place, and after that I know of nothing that will give more relief than the milk-diet. Old cases of asthma, with chronic bronchitis and emphysema look, and I suppose they feel, like the most miserable people in the world. But there is always a great improvement on the milk cure, especially if they break away from their depressant medicines,—a thing they are very loathe to do.

I made no distinction between the different forms of asthma, as usually classified: Cardiac, renal, peptic, thymic, nocturnal or various forms of hay fever; the greatest possible benefit for all of them is obtained on the milk diet.

CHAPTER VIII.

RHEUMATISM.

The milk diet treatment can be applied to rheumatism, with the greatest confidence in a successful outcome. I have never seen nor heard of a return of any manifestation of rheumatism in any one who had taken the milk cure. It may well be asked: Why are there so many suffering from rheumatism if such a simple thing will cure them? There are several answers to this question, but there is no good reason. Many people with rheumatism, gout, chronic bronchitis, and similar diseases are fleshy, plethoric and over-weight, and the milk diet does not appeal to them because, if correctly taken, it means an increase in weight, at first, anyway.

Very many of them are gross eaters; some hygienic writers go so far as to say that they all are. Not many are willing to give up the pleasures of the table for an exclusive milk diet.

A fast is often beneficial to this class, but most of them will not listen to it. If they fast for a time they commit such excesses when they resume eating as to nullify all the benefits.

Those who can afford it go from one hot spring to another, drinking vile tasting and smelling waters, taking mud baths and being doctored by all kinds of quacks, with all sorts of medicines. The resorts where a "good table" is set appeal to them the most and hold them the longest. Very few permanent cures are performed in these places.

Some, like the Arkansas Hot Springs, often greatly benefit invalids, but there the water is quite pure, and the good results come from the change to the pleasant mountain air, drinking large quantities of water, taking a daily warm bath, and experiencing a general improvement of hygienic conditions.

I have seen fully as good results in Long Beach, California, from a free use of the soft, artesian water, together with a more correct diet.

The milk cure does not meet with the ideas of most rheumatic people because they must take a fast to begin with, and next they must cut off their meat, eggs, tea, coffee, whiskey and tobacco.

They cannot see the need of going to bed for they think that would make them weak; they would rather hobble around for years like cripples than go to bed for a fortnight and get well.

Many sufferers from rheumatism have started in bravely on the milk diet, but have stopped short when a natural reaction occurred. In this disease, as in many others, the first sign of a cure is a stirring up of the old trouble, causing often a recurrence of the rheumatic attacks.

It is very common, almost the rule, I should say, for a case of chronic rheumatism, starting on the milk diet properly, to have a return of the old malady. If the disease has prevously taken the form of lumbago, after a few days of the diet, some movement, or muscular effort, will suddenly bring on a typical attack of the pain and spasm. If the patient keeps on with the milk the attack disappears after a day or two, but within a few days more a second attack may come on, but always much lighter than the first. I have seen even a third attack, but so slight as to cause no inconvenience.

If the patient goes through the first attack without ceasing the regular taking of the milk, any further appearance of the trouble will not hinder a cure, because it will be evident from the lighter form that the disease is being mastered. The explanation of these "crises" may be found in the fact that the circulation of the blood is greatly stimulated while it is not yet purified. The excess of fibrin, the uric or lactic acid, or whatever the rheumatic poison may be, is still in the blood and being driven around with greater force, or into parts where previously the circulation had been stagnant, it is only natural that such reactions should occur. These things are discouraging to people who have been in the habit of taking medicine to relieve the attacks, and who have considered that medicine the best which most completely and quickly stopped the pain and discomfort.

It is the old story of something quick and easy, some immediate effect, the suppression of some symptom which is only the surface indication of deeper trouble; present relief regardless of future trouble.

The process of eliminating the rheumatic poison on a milk diet, if slow, is sure. Milk does not contain the elements from which this poison is made, and gradually, the new blood, working within the body, assisted by the baths keeping the external skin soft and porous, drives rheumatism out of every tissue of the body, and, best of all, there is a complete correction of the abnormal process of assimilation, digestion, or elimination which allows this poison to accumulate in the blood.

It is a genuine cure, not simply temporary relief. We do not know why certain people should have rheumatism, when others, living apparently in the same way, should be free from it; why one class of people is subject to rheumatism, but never have tuberculosis, and another class is liable to be consumptive, but never rheumatic. Innumerable volumes have been written on rheumatism, gout, Bright's disease, chronic bronchitis, asthma, etc., but to my mind they are all different manifestations of the same disease, and that disease is simply the deranged condition of the assimilative or eliminative organs which permits the poison to accumulate in the blood.

Undoubtedly the class of food eaten has

an important influence, for meat extract, meat and fish always increase the amount of acid in the blood, while a vegetarian diet always decreases it. Tea, coffee, and cocoa also contain the so-called purin bodies or bases, of which group uric acid is a member. Milk is absolutely free from these bodies.

Prof. Weir Mitchell says uric acid disappears from the urine while skimmed milk alone is being taken, but reappears on the addition of other foods, especially meat. I do not think skimming the cream from the milk would make any difference. I am sure that my rheumatic patients obtain freedom from uric acid by the use of unskimmed milk. The urine of these patients from being highly acid, changes very quickly to the normal condition of a very slight acidity, due to acid phosphates. The perspiration, however, continues highly acid for several days, and sometimes for weeks. The odor from a rheumatic patient on the milk diet is distinctive and unmistakable, but becomes gradually less as they go on to a cure. Many other patients have odors of more or less intensity emanating from the skin, and none of them may be considered cured as long as this persists, no matter how well they may otherwise appear.

There is no form of rheumatism, acute or chronic, of the bones or muscles, so far as I know, that cannot be successfully treated by the milk diet. But there are some cases with obscure, deep-seated pains, probably in the bones and worse at night, which are not rheumatism at all, and cannot be cured in as short a period as rheumatism can.

I have treated several cases of rheumatoid arthritis with bad ankylosed or stiffened joints. The progress of the disease always stops on the milk diet, and, to my surprise, at least two patients recovered movement in joints which I had thought permanently stiff, but the diet was continued for some months in both cases.

CHAPTER IX.

VARIOUS DISEASES.

I am often asked if the milk diet is good for this or that disease, or if it will suit certain cases, as, for instance, where there is an aversion to milk, or a dilated stomach, or where it causes constipation or diarrhea, or if it is not dangerous to use in heart, or kidney disease.

Anyone can take the milk diet, if he starts right, and it is good for any chronic disease, without exception.

I have never seen it used, in full amount, for acute diseases like typhoid fever, nor would it be practicable to give it during the height of an attack of appendicitis, but I have seen many cases where health was lost through typhoid and restored on the milk diet, and other patients subject to periodical attacks of appendicitis, have remained in perfect health after their course of milk diet.

It makes no difference whether a person likes milk or not. One of my most satisfactory cases was a lady who had not been able to take any milk for over forty years. There is no great difficulty in handling the proper amount of milk if it is taken in the small and frequent doses that I recommend, and the patient kept at rest until the stomach and bowels are working well.

Certain specialists say that a dilated stomach is a bar to an exclusive milk diet. This is a great mistake. A dilated stomach is one of the easiest things to cure on the milk diet. Probably if only a small quantity was taken it might aggravate the trouble, but where the proper amount, or anywhere near it, is given, there is no difficulty in restoring the tone, strength and proper size of a weakened and dilated stomach. I think the stomach is the first organ to be restored to a healthy function, in nearly every case.

Chronic diarrhea can be cured on a milk diet in a few days, if the patient can take sufficient milk to raise the blood pressure and heart pulsations. One lady with a long standing case of this trouble, took milk for four weeks, with little apparent improvement, except that her weight increased from 85 to 94 pounds. She had been weak and nervous, largely, I think, as a result of a severe capital operation she had undergone at Battle Creek, Michigan. She could only take about three quarts daily, and this seemed to be too little to restore the proper circulation. But fortunately, when she resumed ordinary diet her bowels operated in a perfectly normal manner, and after nearly three years her health continues good. A gentleman from Maine who suffered from membranous colitis, which had become so acute that he was losing five pounds a week in weight, took the milk diet carefully under my supervision, and at the end of five weeks I discharged him permanently cured.

Regarding kidney disease, some doctors without personal knowledge on the subject have declared that such excessive quantities of milk would ruin any kidney. In answer I say that the amount is not excessive, but only the natural amount required, and that the kidneys do stand it, and grow healthier every day. It is no hardship for any organ of the body to exercise its natural function when it is given the proper material to work upon. The function of the kidney is to separate from the blood and eliminate from the body certain salts and waste materials, and raising the blood pressure and increasing the amount of blood makes this function easier. The fact that a much larger quantity of urine is passed on the milk diet than usual only indicates that the work of elimination is assisted by the larger amount of water passing through the kidney.

The urine of a healthy person is more or less poisonous. It contains toxic materials eliminated from the blood. In Bright's disease these poisons are reduced in quantity or disapper from the urine because the diseased kidney is no longer capable of separating them from the blood. The sweat glands are able to excrete a portion of this matter through the perspiration, but sooner or later its retention in the blood is apt to cause uremic convulsions.

As the urine is increased to three or four times the usual amount by this treatment either one or both of two things must happen The waste and toxic matter is very much diluted by the additional water, or, a very much larger amount of waste is excreted.

In the first case the elimination should be easier by reason of the greater amount of fluid washing out the tubes; in the second case the blood is purified more rapidly.

The human kidney is never found in a perfect condition. Sections for microscopic study in physiological laboratories are usually made from the kidney of some of the lower animals, or possibly from that of a child. The adult kidney always shows more or less pathological change in the delicate structure and complicated arrangement of the uriniferous tubules.

I have seen many very serious cases of albuminuria and Bright's disease take the milk cure, during the last 27 years, and I have never heard of any of them having any trouble from the disease afterward. I believe that diseased kidneys are restored to almost perfect condition by the milk diet.

Regarding heart disease, meaning usually organic or valvular disease, there is a general impression that the less fluids given the better it is for the patient. Schroth's method of an almost entire dry diet was thought to be good for heart disease; Oertel's method of dry diet, with active exercise, helped many cases, and Tuffnell's treatment of absolute rest with a dry and very restricted diet has made some remarkable cures.

I was, for many years, under the impression that the milk diet could not be given in severe forms of heart disease, but I have so completely changed my views, simply from the result of observation, that I now feel sure the milk diet can be given to any case of heart disease, with the greatest success possible to any treatment. Further than that, I make the prediction that the future treatment of severe forms of organic and functional, or nervous heart disease, and aneurisms, will consist almost entirely of the milk diet and rest.

A broken or ruptured valve may not be restored by the milk diet, or any other treatment, but it can be compensated for, so that the possessor may live in comparative comfort. Almost any other form of heart disease can be successfully treated by means of the milk diet. The simplest case of all is the weak heart of an anemic person. Such people, with poor circulation, white, pasty looking skin, usually underweight, but sometimes fat and flabby; weak and languid, disinclined to exertion and easily tired, with typical heart sounds— the "anemic murmur," and a small, undeveloped heart, can obtain perfect health on milk and rest. I have seen some of them lose the murmur in two or three days, followed by a steady growth of the heart in size and strength, with a corresponding improvement in the general health.

Every doctor sees young people with the disease, or condition, rather, called chlorosis, or "the green sickness," which is often benefited by the administration of iron, because the blood of such people lacks the necessary amount of iron. Milk always contains iron, and a short course of the milk diet always cures chlorosis, and puts the general system in good order, a fact which is not always the case after the administration of medicine.

Some of the more severe forms of heart disease are complicated by dropsy of the feet and ankles, and other parts. Although perhaps of months' standing, this dropsy always disappears in a few days on the milk and rest. Rest alone often relieves these cases, but rest without milk will not cure them.

Regarding dropsy, of any part of the body, from whatever cause, I have never heard of a failure of this treatment to cure it.

Any form of valvular disease may be

treated by the milk diet, with the greatest possible advantage, but in these cases, more than in any others that we are called on to treat, rest, COMPLETE, ABSOLUTE REST, is essential.

Two cases of valvular disease of the heart have lost several pounds weight in the first few days on a full milk diet, but afterwards gained satisfactorily. I mention this fact (which I cannot explain) because it is very rare for any patients (except those having obesity) to lose weight while on the milk.

I have stated elsewhere that the heart increases in size, and this is true of every case, except the dilated weak heart. It is not difficult to prove this, as any patient with a weak heart can observe the apex beat of the heart move downward and to the right as the cure goes on. Growth of the heart follows the law governing all muscles, that, if the nutrition is kept up, increased work is followed by increased size. The very first work that the blood made on a milk diet has to perform is to nourish the heart itself, for the first arteries leading from the aorta, or main artery of the heart, are the coronary arteries, which turn back into the heart muscle to supply it with blood.

Right here I wish to speak of a very common, but serious, disorder, the treatment of which by the milk diet is so successful that the fact should be universally known. I refer to hardening of the arteries, or ARTERIO-CLEROSIS, sometimes spoken of as a physiological process.

"A man is only as old as his arteries," refers to the fact that, while hard arteries are a frequent accompaniment of old age, they may also be present in a young, or middle-aged man, who has lived improperly.

Hardening of the arteries is perhaps the first apparent change in the blood vessels that indicates beginning degeneration. It is included by medical writers in the descriptions of degenerations, and, while it is admitted that the condition may continue for years before it becomes dangerous, it is generally considered the beginning of the end. It is surprising to see how little hope is held out to persons afflicted with this disease, by writers on the subject, who seem to take it for granted that there is no cure, nothing to do but to make them as comfortable as possible, "enjoin them to lead a quiet, well-regulated life, avoiding excesses in food and drink." "It is usually best to frankly explain the conditions of affairs," etc.

After a certain time, the duration of which varies in different cases, the inner lining of these hardened arteries softens, ulcerates, and breaks down, resulting in aneurism, embolism, paralysis, and apoplexy.

I believe that almost every case of this disorder can be cured, if treatment is used before the ulcerative stage begins. One old gentleman whose pulse felt like a wire, took the milk diet treatment three years ago, and in less than four weeks his arteries were soft, and the neuralgia of the heart, from which he suffered intensely, disappeared. Neither the sclerosis nor the heart trouble has ever returned since that time.

One disease that is difficult to treat with a milk diet is chronic inflammation of the bladder, especially the tubercular form. This is usually considered incurable, and I know nothing better for it than this treatment, but these old, inflamed bladders with thickened walls and degenerated linings, have a very small capacity, so that even on ordinary diets urination must be performed very often. With the amount of urine increased three or four times, and the diseased organ showing no immediate improvement, it requires considerable faith, or persistence, on the part of the patient to carry on the treatment. The rubber urinals sold by druggists are of material assistance, and as the urine becomes soft and unirritating, the bladder can hold more of it than while the person is on an ordinary diet. Indeed, it is a great relief to these people to have the urine change from the fetid, irritating, ammoniacal, decomposing stuff to the almost colorless and odorless water characteristic of the milk diet.

The milk diet has proved invaluable to many young women with painful menstruation, misplaced, or undeveloped wombs, and other disorders peculiar to the sex. Fibroid tumors of the uterus have been known to disappear under this treatment, and many other conditions, apparently only suitable for operations, have been remedied. There is a great development of the pelvic organs, while resting and taking the milk diet.

Usually the best time to start in with the milk is right after a menstrual period. The next period may be, and often is, irregular as to time, but if three weeks elapse before it comes on, it is almost always the case that the pain will be greater than usual, and some patients are inclined to stop the milk to get relief. The pain is due to the unusual amount of blood, in excess of the ordinary congestion at this period. If a woman can stand the pain, and keep the milk going, greatly improved conditions will be noticed at the next menstruation.

The relief experienced by these cases is similar to that occurring after a normal childbirth, when, as most married women are aware, menstruation almost always ceases to be painful.

There is, perhaps, no part of the body that receives greater and more uniform improvement in this treatment than the generative organs. Milk is the best nerve food that we possess, and the connection between the nervous system and the sexual organs is a very close one. The improvements in the two go on coincidentally.

An exclusive milk diet has been in use for patients with diabetes mellitus since 1868, when it was introduced by Donkin. Prof. Tyson tells of one case taking fourteen pints daily. He regards the diet as the most important part of the treatment of this ordinarily fatal disease, and is confident that the exclusive milk diet is the most effectual way of treating it

I should feel certain of good results from the milk diet treatment in diabetes, but I regret that I have no cases to report, never having succeeded in getting a suitable one for a trial.

Since the above was written seveal diabetics have attempted to take the milk diet, but none have carried it on more than a week. They all lost weight, but two gentlemen seemed benefited by the short course. I still believe the treatment would cure suitable cases, not too far advanced.

I receive numerous inquiries regarding cancer, but I can only repeat what I have just said about diabetes, as I have the same lack of experience with true cancer.

CHAPTER X.

PSYCHOLOGY OF THE MILK CURE.

I am glad to say there is no "suggestion," or faith cure about the milk diet treatment. You put something in, and you get something in return for it every time. I have had patients who took the treatment because friends urged them to, but without the least faith in it, except that they thought "milk wouldn't hurt them," and these people have made as good a cure as others who had perfect confidence in the method.

All my patients will bear me out in the statement that no psychological influence has been exerted. The milk diet treatment is a simple thing, and within any person's reach. It is not necessary to have grand buildings, expensively furnished apartments, showy bathrooms, glittering apparatus, or complicated appliances of any kind. What is needed is a quiet, cool and airy place, with a comfortable bed and the necessary toilet utensils. A vase of fragrant flowers is always acceptable, but showy ornaments or pictures, or lace curtains, are out of place.

Nor is it necessary to have frequent examinations of the body, of the urine, the blood, or the secretions of the stomach, etc. After a long sanitarium experience, and listening to the histories of many people who had been the rounds of the various institutions, I am firmly of the opinion that these "examinations," as usually conducted, are mainly beneficial to the staff of young doctors who get the fees, and incidentally, some experience. Of what possible use is it for the patient to learn one week that he has "hypoacidity," and the next week that he has "hyperacidity," if his dyspepsia is not relieved?

Such things may have their use in sanitariums conducted with the idea of having the patient stay as long as their money holds out, but they are not needed in a place where the treatment does what it is claimed to do, and patients are steadily improving.

The average person will get better results in a well conducted sanitarium than he will at home, not especially on account of more skillful treatment, but from causes that are well understood by all physicians.

The change of air and scene, the making a regular business of the "cure," the relief from home cares and worries, the getting away from the well-meant but often harmful solicitations of anxious relatives and friends, often the exchange of a stuffy, over-furnished, over-curtained, badly ventilated bedroom for a more healthful one, all these, and many other details, frequently assist in getting an invalid started on the upgrade. But more than all other things combined is the wonderful influence of the new blood made so freely on the milk. Rest and quiet, daily warm baths, and plenty of fresh, pure air are necessary to most people in order that they may take and assimilate the proper amount of milk, and eliminate the waste products.

It is not necessary to have a daily "Health lecture," during this treatment, but frequently, at the start, a little encouragement is helpful to keep the milk going down, because the senses of hunger and thirst do not cry for it, and it is easy to stop drinking for a while.

The best "cures," in my experience, have been the patients who started in with the full amount of milk, and took it continuously, without interruption other than during the sleeping hours. They did not stop because their stomachs seemed full or for a bad taste in their mouth. Some of them have disregarded nausea and even vomiting during the first days of their treatment. Others have suffered headaches and backaches, and, later on, the dull, stretching pain in the stomach, kidneys and liver, which may accompany the rapid growth of those organs. Very many have had returns of the old pains of rheumatism, neuralgia, earache, toothache, pleurisy, peritonitis and inflammatory conditions of the generative organs, which they may have had years before. These pains usually last about a day, but in chronic cases of long standing, where there has been considerable growth of inflammatory tissue, and adhesions, as between the serous surfaces of the peritoneum and various organs of the abdomen and pelvis, the duration of the pain is somewhat in proportion to the length and seriousness of the disease.

The pain is never as severe as it was in the original disease, except perhaps in some women at the menstrual period, and the pain may be stopped by stopping the milk, and thereby taking off some of the blood pressure, but that is usually the wrong thing to do, for it is the pressure of the excess of blood that works the cure.

I do not ask the impossible of any person, but I tell those who are inclined to stop the milk that the pain is only a necessary reaction in the diseased part; that pain means a growth of new capillary bloodvessels in a place where the circulation had been stagnant, that the part, or organ is growing larger, getting straightened out, coming back to the place where it belongs, stretching its fibrous and sensitive covering, (as in the liver and kidneys) pulling on the contracted ligaments, or abnormal fibrous bands which have bound it down and interfered with its action, or stimulating the normal movement where it had been paralyzed.

It is easier to understand why there should be pain with a curative process than it is to explain how there ever can be a cure without pain.

When I have told patients what, to the best of my knowledge and experience, is going on, I leave it to them to decide whether they can stand the pain, or discomfort, with the expectation of a complete cure, or whether they will have to stop the milk temporarily, and perhaps stop the curative process when it is at its height.

I encourage them by stating the fact, which I cannot emphasize too strongly, and which everyone should remember, that in 27 years' experience with this treatment, on all classes of patients, suffering from heart disease, brain and nerve disorders, blood clots, paralysis, inflammation of the bowels, ulcerative and hemorrhagic processes in various parts of the body, chronic specific disease, or chronic poisoning due to lead, mercury, arsenic, or any medicine, I have NEVER HEARD OF ANY INJURY, OR BAD RESULTS, from pushing the milk diet TO THE UTMOST CAPACITY OF THE PATIENT.

The one possible exception to this rule is the case of a man who had a blood tumor, or aneurism, resulting from a recent gunshot wound, which burst from the pressure generated by a milk diet, with fatal results.

I do not know whether the man was resting, or how much milk he was taking, or any other particulars, but I would advise against giving the full milk diet to any patient who had recently been operated on, or who had a ruptured artery from any cause, except chronic disease. I am not afraid to give the milk diet in any case of diseased bloodvessels, or in aneurism caused by disease, for I believe the blood carries its own cure for these conditions, and the more of it the better, but COMPLETE REST MUST GO WITH IT.

It is not possible in this little book to follow each case to the end. There is an infinite variety. If you have learned the great natural principles upon which the treatment is based and follow the directions I have given, you will be ready for any condition which may arise.

Don't leave out some portion which you think is unnecessary, nor add something to it which has helped you under other circumstances. Try my way first.

CHAPTER XI.

REST.

There are a number of reasons why complete rest must be had, at least during the first part of this treatment. One very practical reason is the fact that many weak stomachs cannot retain the milk unless the body is lying quietly and therefore more or less relaxed. A stomach that has long been making an insufficient supply of blood is in a rut, and is disinclined to take more food, and thereby be compelled to make more blood. Practical experience has shown that if the body, (and stomach) is kept as motionless as possible, the necessary amount of milk is much easier retained in the stomach and digested.

The same principle holds good on a sea voyage. All old travellers know that lying down at full length in the berth until used to the motion of the vessel often prevents seasickness.

Another reason is that naturally, in all animals, digestion and assimilation go on better while the animal is at rest, or asleep.

But the most important reason for resting while taking the milk diet may be explained as follows: The treatment is taken to correct some function, or to develop some part of the body; something is wrong, or lacking, or needs rebuilding. In short, growth is necessary, and growth is always a function of rest. We may, by exercise, build up big muscles, but the growth even of muscles, is performed between the periods of activity, for work always uses up energy and wears out cells. Continuous work, without relaxation, would be impossible for muscles or other tissues. The intervals of rest between the periods of work enable the blood to flow freely into the part and carry the needed nourishment to replenish the cells exhausted by the previous energy. Work may be the stimulant which causes subsequent growth, but in itself work is exhausting, destructive. Recovery and recuperation can only occur during relaxation; **we grow while resting.**

The body requires its night's rest after its day's work, and for the same reason a body weakened by a long period of strain, misuse, illness, must have a period of rest, in some measure proportionate to the period of wear. If, during this period of rest, there is an increased supply of nutrition and blood, we have the ideal conditions for rapid repair. With the wear and tear and waste of the muscular system stopped, the nervous energy which usually directs it is saved, or diverted to more useful purposes. The voluntary muscles are useful as organs of locomotion, prehension, etc., but they are not vital organs. Men have lived minus all four limbs.

In chronic illness it is the vital organs that we have to deal with, those concerned with digestion, nutrition, respiration, circulation, innervation, and depuration. By putting at complete rest as many of the muscles as may be possible, we save a large amount of nourishment and nerve force that would otherwise be expended without any useful return.

Every unnecessary drain must be stopped to allow the vital organs to rebuild and restore themselves. The more complete the inactivity of the external muscles, the brain and nervous system, the sexual organs, the better the prospect of restoring the normal function of the other organs provided plenty of blood is supplied.

I deem it an unfortunate, but unavoidable feature of the treatment that the organs of generation almost immediately share in the general improvement, because it is undesirable, at this time, to spare any of the blood from the important work of reconstructing the digestive apparatus, and the lungs, (if there is a pulmonary disease,) and there are too many men who cannot restrain themselves.

Some people are unhappy with a few dollars in their pockets and won't be satisfied until it is spent, instead of putting it in the bank and accumulating a good working surplus. I hope this simile will be understood and appreciated by married folks, and others.

The success of the Weir Mitchell treatment is largely due to the complete rest prescribed for severe cases. For weeks these patients are not permitted to sit up, or sew, or write, or read. They are even fed by a nurse, and talking is prohibited.

Complete rest on an ordinary diet usually means that massage will be required to move the bowels, but on the milk diet this is unnecessary.

My patients usually are allowed to read if there are no headaches, and the stomach is taking the milk without difficulty. But the reading should not be continuous. Read for ten minutes between drinks, and then lay the book or paper down for fifteen or twenty minutes. Reading helps to pass away the time, and satisfies people who, without it, would want to be doing something more harmful. But read as little as possible, and never by artificial light.

Talking is usually unnecessary and seldom beneficial.

Don't think because you are lying abed for weeks and keeping quiet that you will get rusty. I never knew the rest part of the treatment to do any damage; most of the patients are inclined to get up too soon, rather than stay abed too long. But they all store up energy while resting and the good effect is apparent as soon as they return to ordinary life.

Many people with tired nerves and poor stomachs cannot take a sufficient quantity of milk to do much good without being completely relaxed. But this state of relaxation is a hard one for some people to get into. They don't want to go to bed, and when they do, they stack up pillows behind their backs, until they are almost in a sitting position.

They are losing half the benefits of the treatment, and the opportunity of a lifetime to take a complete rest. Isn't it worth while to really rest for a few weeks if comparative comfort can thereby be secured for all the remaining years of life?

To enable these folks to let go a little, to reduce the tension, the warm bath is of great use. In the warm bath only, do some of them first learn to relax. It is sufficient for some people, to tell them to lie out flat in bed, breathe deeply a few times, and then, beginning with the head and neck, relax all the muscles of the body, so that if the various parts were lifted they would fall like logs of wood. When all the muscles are relaxed there is a pleasant sensation, almost like floating in the air. Sleep secured after getting in this state is far more restful than where one simply drops off from fatigue, with all the weight of the day's work and cares distorting the body.

But many cannot properly relax at first. Here comes in the benefit of the warm bath. It is not "weakening" for these strained, nervous cases, any more than sleep is, but it does permit them to relax. Nothing supports the whole body so gently and easily as a good tub bath. I notice the insane asylums have grasped the idea, and many of them are fitting up bathrooms where very nervous cases may remain continuously in the neutral bath **for weeks at a time,** eating and sleeping therein until the nervous system has recovered.

The relaxing and soothing effect of the warm bath is due to several causes. It is sufficient to mention here the warmth, which relieves the body of its heat generating function; the moisture, which is absorbed externally and inhaled internally; the cleansing and opening of the pores of the skin; the softening and removal of the dead epithelial scales; the growth of new capillaries; the relief of pain and soreness and the wonderful buoyancy caused by the equalization of the pressure on the surface of the body. No cabinet, or vapor bath, or electric light bath can do what the warm tub bath does in combination with the milk diet.

When the patient has learned how to relax the body, and really rest, I have little doubt as to the final result of the treatment.

To illustrate the great difference in taking the milk diet, with and without rest, I obtained permission to quote the following case:

Mr. Aubrey Parks of Omaha, Nebraska, was attacked by acute nephritis, or Bright's disease, about three years ago. It ran on for several months and finally became chronic, with a great deal of dropsy, in spite of treatment in two hospitals and by several good physicians. He finally went to a sanatorium in Michigan where the treatment consisted of a long fast, followed by an exclusive milk diet, a glass at a time, at frequent intervals, as I recommend. But, instead of resting, he was ordered to exercise daily, and went to the milk room every half hour for his milk.

The result was that while his dropsy and albuminuria decreased somewhat on the fast, both increased markedly as soon as he started the milk diet. He was ordered to take another fast of about two weeks and then again took the milk diet, with no better results than before.

Shortly afterward he wrote me about his case, without informing me, however, that he was not resting while drinking milk. I replied that I could not understand it, as I had never had a case of dropsy that was not cured on the milk diet. Mr. Parks finally made the long trip to California to take the treatment in the manner I recommend.

On his arrival here September 1st, 1909, he showed a condition of general anasarca, or dropsy, literally all over the body. He could not wear any of his regular clothing, hat, or shoes, on account of the swollen, water-logged condition of his skin. His weight was 186½ stripped, although he had been fasting several days during his journey. By my direction, Mr. Parks went to bed and remained there over a month, except for the time he spent daily in a warm water bath.

He took from six to seven quarts of milk in twenty-four hours, and passed some days over ten quarts of urine. In fifteen days his weight had gone down to 127—a loss of almost 60 pounds.

From that time he slowly gained weight, up to 154 pounds of solid flesh, although the dropsy did not entirely disappear for several weeks, the ankles being the last to become normal. The albumin in the urine persisted for nearly the two months he was under my care, but finally disappeared. Mr. Parks, sixteen months after this treatment, is living in Long Beach and is quite well.

No case that I remember shows so emphatically as this one does, the great benefit of rest while on the milk diet.

Another case almost as instructive is that of Mr. S.—of Iowa, who being attacked by a slight stroke of apoplexy, went to the same sanatorium in Michigan of which Mr. Parks had been an inmate. Mr. S. knew that his arteries were in a diseased condition and this condition had no doubt caused the ruptured artery in the brain.

He took the usual fast for about two weeks and then started in drinking milk, exercising vigorously every day, according to the system in vogue there. In less than forty-eight hours he suffered a second stroke which paralyzed his right arm and affected his speech,—a result I should have expected under the circumstances, as the fast could in no way have strengthened his blood vessels to withstand the blood pressure consequent to exercise on a milk diet.

This man came to me as soon as he was able to travel, in January, 1909, and after a short fast he went to bed and took five and a half quarts of milk daily for four weeks. I never had the slightest fear of another hemorrhage, because he was not making any exertion that could be avoided.

After four weeks of rest and milk diet, I felt confident his arteries were in condition to stand exercise and gradually he began walking and using his arms. In less than a week, he could walk over two miles at a time, and soon after returned to his home. He wrote a few months afterward that he was resuming his occupation as a traveling salesman, and felt well.

CHAPTER XII.

EXERCISE.

A course of milk diet alone does not insure a person forever against future illness or the encroachment of old age, but it certainly tends to do so when combined with right living, and by right living I mean correct eating, drinking, working, thinking, sleeping, breathing and exercising. Of all these, the last is by no means the least; for many people it is the most important of all.

There are many useful books and periodicals devoted to exercise and physical culture, and I advise all my patients to keep up their interest in the subject by studying them, for it is easy to drop back into the old habits of inactivity and weakness. I cannot fully cover the subject in this work but there are some points that I will speak of, not always mentioned in exercise instructions.

THE OBJECT of all exercise should be to improve the circulation of the blood, increase the breathing power, and develop organic vigor in those vital parts of the body on which our wellbeing depends.

Exercise simply to build up big muscles and to do imposing feats of strength is largely a waste of vitality.

In exercising any muscle, we also exercise the brain and nerves, because they control the muscles, and when we get tired, it is because the nervous apparatus is tired, not the muscle itself.

When we decide to make a muscular movement, the brain sends an impulse over the nerve or nerves extending to the special muscles involved, ordering the muscles to contract. As a muscle in contracting uses up a certain amount of food carried by the blood, the artery supplying blood to the particular set of muscles in use becomes enlarged or dilated in order that the necessary food and oxygen may reach the part. This itself is a muscular act, the little involuntary muscles in the walls of the artery being controlled by the sympathetic system of nerves. After the muscle has been working, the vein leading from it to the heart carries a larger volume of blood containing the waste matter created by the muscular effort. The brain itself requires more blood when engaged in manipulating the muscles, just as it does when engaged in any purely mental process, for tissue cells in any part of the body cannot be active without necessitating a supply of blood in excess of the amount ordinarily sufficient to nourish the part.

So in muscular exercise there must be also nervous exercise. Just as the muscle grows larger, stronger, and firmer by exercise, so the brain and nerve cells become more efficient in the way of rapidity, acuteness, and precision in the execution of the movements. And muscular effort, by throwing into activity a different portion of the cerebrum, and causing a greater flow of blood in perhaps little used channels, may relieve conditions like worry, anxiety, insomnia, nervous exhaustion, and even pain.

The most important effect of muscular exercise is to increase the number and depth of respirations, and thereby the quantity of air passing in and out of the lungs, leading to an increased absorption of oxygen, and elimination of carbonic acid. It is estimated that a man at rest draws in 480 cubic inches of air per minute; if he walks four miles an hour, he draws in five times as much, or 2400 cubic inches; if he walks six miles an hour, he draws in seven times as much, or 3360 inches.

The effect of exercise on the skin is to dilate the cutaneous blood vessels, and increase the amount of perspiration, thereby allowing more water, salts, and acids to pass from the system. The evaporation reduces the temperature of the body, which would tend to rise.

Exercise increases the appetite, doubtless as the result of wear and tear of the muscles and other organs. Digestion is more perfectly performed, and the circulation through the liver and portal system is quickened. There are many different systems of exercise and probably they all have good points, but whatever system is used, especially where no apparatus is required, it must be kept in mind that to be of any benefit real exercise must be taken. That is, the muscles must be firmly contracted, and the brain kept occupied with the matter in hand. It will be of little use to flop the hands back and forth in a languid manner, while the eyes are gazing out of the window, and the brain making plans for the day's work, or something else.

Throw energy and concentration into the movements; they need not necessarily be made rapidly, although rapidity is a factor in some exercises, but contract the muscles to the utmost limit. The alternate contraction and relaxation of the muscles is what drives the old blood out and pulls in new, and the greater the concentration the more efficient the movement. Make the limb or portion of the body being exercised as rigid as possible. In the arm movements, especially the overhead ones (and these are particularly useful because bringing into play muscles that are otherwise little used) leave the hands open so that the muscles of the forearm will not be much affected. These movements should be concentrated in the upper arm, shoulder, neck, chest and upper back. Taken in this manner, they are of the utmost benefit to the chest, neck and spine. A few moments' exercise of the arms, in this manner stimulates the circulation in the great blood vessels in the neck, and along the spinal column into the brain, and particularly brings into activity the great nerve centers which lie along this tract.

After patients have finished the resting part of the treatment, they dress and sit up an hour or two the first day. This is usually enough where patients have been in bed several weeks, because the muscles, although big and firm, are unused to exercise, and they tire easily the first day.

As soon as patients begin to be up and around, the length of the bath is decreased a little every day, until only a short, 10 or 15 minute, bath is taken. Instead of finishing up with the water quite warm, no more hot water is added, and, when the short bath is reached, a cold sponge bath, or shower, can be used after the warm bath.

The second day the patient is up, a short walk can be taken, and every day afterward the exercises are increased, and the time in bed decreased, until only the necessary time for sleeping is spent in bed.

Two or three days after stopping the complete rest, start in on the following exercises and take them every morning and night. They should be done standing in front of a mirror, with very little or no clothing on. The exercises are simple, and few in number, so as to be easily remembered, and not take too long in execution. It is better to do a few exercises twice a day regularly at home ,than to go to a gymnasium two or three times a week and work hard for an hour or two.

All these exercises can be made as hard as desired by simply increasing the tension, or rigidity of the muscles.

EXERCISE NO. 1.—Stand upright before the mirror, arms down, palms facing forward. Breathe deeply and regularly. Make the muscles of arms and shoulders as rigid as possible. Flex the arms at the elbows alternately and bring the hand up in front of shoulder. As one hand comes up the other goes down. After doing each up and down movement relax completely all the muscles of that side. Do the movements slowly, starting with five for each arm and increasing one a day until ten is reached.

EXERCISE NO. 2.—Stand upright, arms extended sideways as far out as possible, palms up. Make as rigid as possible all the muscles of arms, shoulders and neck. Bend both arms at the elbows, until hands are near face, then extend to starting position and relax. Do this the same number of times as Ex. No. 1.

EXERCISE NO. 3.—Stand upright, arms extended sideways as far as possible, palms facing forward. Make rigid muscles of arms, shoulders, back and neck. Bring hands forward until almost touching in front of face, and while doing so exhale, or empty the lungs as completely as possible. Then return hands to starting position slowly, inhaling as you do so until lungs are completely filled. Then relax. Always inhale through nose with mouth closed. Repeat same number of times as Ex. No. 1.

EXERCISE NO. 4.—Stand upright, arms at sides, palms inward. Make rigid muscles of arms, shoulders, back, sides and abdomen. Bend body at waist line sideways to left side as far as possible. At the same time bend right elbow and bring hand up into armpit, closing fingers. Relax momentarily and straighten up. Repeat to other side. To be done five times.

EXERCISE NO. 5.—Stand upright, arms extended sideways, palms up. Make arms and neck as rigid as possible, and raise hands straight over head, without bending elbows. Return to starting position and relax. Repeat same number of times as Ex. No. 1.

EXERCISE NO. 6.—Stand upright and raise arms sideways until hands are above head. Then bend over forward until finger tips touch floor as far as possible from feet. Gradually straighten up, raising hands with palms facing toward body. As you assume perpendicular, rotate hands outward until palms are facing forward.

This like No. 3, should be made a breathing exercise, exhaling as the body is bent forward, beginning to inhale as body comes up, and expanding chest as much as possible at the moment you rotate hands out. When the expansion is as complete as possible, hold the breath as you raise arms sideways as far as possible, and commence exhaling as body bends over, etc. To be done slowly five times, allowing 8 to 10 seconds to each complete movement.

I have given principally arm exercises, because it is the upper part of the body that stands in greatest need of development, and these movements markedly help the breathing, and also are beneficial to the brain, and nervous system in general.

The body bending exercises are particularly useful in constipation, but should be done by everyone. All these movements assist in maintaining the poise of the body, and tend to gracefulness and agility in men or women; old or young.

The man of 45 would be in greater demand by employers if he would keep up his strength and quickness, and brush the cobwebs out of his brain by vigorous daily exercise. He who is slow of speech and movement and unwilling or unable to stand an occasional "rush" in his work, will likely be superseded by a more alert and active employee.

Age is no drawback, but inefficiency is.

For the lower limbs, walking is good exercise, when properly done, with the head up, the chest expanded, and the limbs swung vigorously and not dragged along. Running is still better and helps immensely to improve the lung capacity. If outdoor running cannot be practiced, a good substitute may be found in the room by dressing in a light costume, or none at all, and going through the motions of running, but dropping the feet in the same place all the time. Rope skipping is another strenuous exercise that can be practiced in a room.

Handball is excellent, but if you have no

court get a light, hollow rubber ball which you can throw up in the corner of your room, near the ceiling. It is perfectly noiseless, often a valuable consideration, and the return direction is so variable that it keeps every muscle on the jump to catch it.

Every year I am more convinced of the benefits of exercise after taking the milk diet treatment. The patients who remain in the best condition are those who systematically exercise.

A simple test that can be made by any one is to perform, night and morning, the tensing exercises as given in this book for, say, ten days, noting carefully the general condition every day, as to strength, activity, appetite, bowel movements, mental condition, etc., and then stop them entirely.

After another ten days, resume the exercises, and I am sure, on comparing your condition, with and without exercise, that you will decide in favor of keeping up the exercises.

CHAPTER XIII.

AFTER TREATMENT.

Almost all patients who have taken the milk diet under my personal direction, have asked me what they should eat afterward; how they can be sure of holding the benefits gained on milk.

The permanent results of this treatment are invaluable, and I believe better than those obtained by any other method, and permanent cures are what is wanted in every case, and not simply temporary improvement while taking the treatment.

While there is a great apparent improvement in all patients taking the diet, yet it is a fact that many of them often have to wait until ordinary habits and diet are resumed before they realize the full benefit gained.

All persons taking the milk cure properly will find their physical condition better, perhaps, than it ever was before. The circulation is active, digestion perfect, and all the functions of the body working well. There is no good reason why these satisfactory conditions should not be retained on a return to ordinary life.

It is impossible to give one set of rules that will fit every case, on account of variations in the individual, such as age, habits, condition of teeth, financial circumstances, location, previous disease, etc., but I can give here certain general directions which will no doubt be helpful, comprising as they do the observation of this class of patients for twenty-seven years.

Most people have some idea of what caused their previous lapse from health, and they should avoid the former errors.

One very common source of ill health is improper breathing, and breathing impure air, day or night. This condition must not be returned to.

Another very important thing to avoid is heavy clothing. No matter how good the circulation, or how perfect the regulation of the body heat, if thick, heavy, tight fitting undergarments are resumed, there is little chance that the natural animal heat will continue to be generated as freely as before. The warm clothing obviates the necessity of producing heat within the body, and the oxidation of the blood becomes less perfect because it is not necessary to use, or burn so much oxygen. The stomach does not make as much blood, therefore cannot digest as much food. The food, if not digested, becomes a tax on the system, the appetite is interfered with, and the general vitality lowered.

On the other hand, with light clothing, and open mesh underwear, the circulation in the skin will be more active and assist in retaining the body heat; oxidation and metabolism in the body will be promoted; a larger amount of food assimilated, and more air respired.

People will put on clothing, and more of it, every time there is the slightest suspicion of being cold, or even if the skin becomes cool, not appreciating the fact that cold is one of the best stimulants we have for the circulation in the skin. And, while quick to add clothing, they are slow to quit it, and when the weather warms up again they have not only missed the stimulus of the cold air, but now are enervated by the superfluous heat of their garments.

Wear light garments and keep the skin in a state of activity. Make your own heat by taking in plenty of oxygen and food and using it as nature intended. If you have to move a little quicker, or breathe a little deeper, to get the necessary oxidation and animal heat, all the better. Your muscles will work better, and even your brain will be more active, if your vital processes are not smothered under heavy clothing.

Notice a man with cold hands, or feet, and see if his limbs are not swathed with garments three or four layers deep. How can you expect sufficient blood to get down to the extremities if it has to go through several feet of flesh that is already too warm?

If the man with cold hands, instead of wearing close, heavy, undergarments, with tightly knitted wristbands, will wear sleeveless undershirts, his hands will be warm as soon as the system is accustomed to the change, and his general health better.

Keep the skin active by frequent baths. Friction baths, using a dry, coarse towel, or a brush, are excellent, and can be taken every morning. A warm water bath is necessary once or twice a week, and a daily sponge bath, or rubdown of all the body usually covered by clothing, helps to keep the skin in order.

An important aid to health in many cases is a correctly fitted pair of glasses or spectacles. It is wise to have the eyes tested after taking this treatment, because the eyes undergo changes as well as everything else, and glasses that were used previously may cause eye strain afterward on account of being too strong.

A plentiful supply of water must be taken daily, and the time to begin the water drinking is right after you stop drinking milk. Take a glass the first thing in the morning, and the last thing at night, and two or three drinks between meals. It is best not to drink any fluid with the meals as it tends to wash the food down before thoroughly masticated. and also stops the secretion of saliva, and stops the starch-digesting action of that already secreted.

If there is any tendency to constipation the water drinking will generally prevent it, and especially, if a little fruit juice is added to the water. Lemonade is a good drink, but the "bracers" dispensed at the soda water fountains should be avoided. Ice cold drinks of any kind are a poor thing to put in the stom ach, and equally detrimental are the very hot drinks that many people take.

When it comes to the question of food I am inclined to be rather liberal in my views, and do not lay down hard and fast rules, except, perhaps, for the former rheumatic patients, where I advise against a resumption of meat eating on general principles. But some of them are eating meat occasionally, and none of them have had rheumatism since taking the treatment.

Many of the patients are vegetarians, but some of them believed in that before taking the treatment, and still lacked health.

Of late years the patients who have been using a diet of mostly natural, or raw foods, have all kept in good health, and I have come to believe more and more in this kind of a diet.

With milk and eggs, cream and cheese, nuts, fruits and vegetables, bread made from whole wheat, rye, oats or corn, and butter, anyone should be able to select a bill of fare.

Cabbage, beets, turnips, carrots, radishes, spinach, green corn, onions, chives, tomatoes, celery, lettuce, and cress are all more digestible raw than when cooked, and have a better flavor. Combinations of these vegetables ground up in a chopper, together with sweet potatoes, or nuts, or fruits with various kinds of dressings composed of olives and olive oil, lemon juice, yolk of eggs, cream or milk, make perfect food dishes, taste good, and are easily digested.

Raw eggs can be taken in several ways, as broken into fruit juice, and swallowed without breaking the yolk, or beaten up with a little seasoning, or as egg lemonade, or stirred up with hot milk, or just placed in hot water a few moments to warm it through, and then eaten as a boiled egg would be. Raw eggs are undoubtedly more easily digested and assimilated than cooked eggs, and I believe they afford more nutriment and energy.

There are some persons, even after taking the milk diet, who cannot use milk as a drink with other foods without noticing a costive effect, but they all, I believe, can take milk alone, or milk and fruit, or bread and milk, and make a satisfactory and easily digested meal. Bread and milk suppers are a favorite with some of my former patients, who are past 80, and have no teeth. Bread and milk and fresh apples make a good combination. Any bread to be used in milk should be slightly stale, never fresh and soft. Whole wheat or Graham bread is far better than the white bread made from patent bolted flour, which contains little but the starchy part of the grain.

Some people like toast, or zwieback, probably because it tastes crisp in the mouth and affords something to chew, but toast is a very poor article of food and usually constipating. It is really nothing but a cinder of carbonized starch and gluten. The day of toast and tea for invalids is gone, and will never return, in sensible households.

The legumes, beans, peas, lentils and peanuts afford valuable nitrogenous food. Green peas are excellent ground up, seasoned and slightly heated. Adding milk or butter improves them. Raw peanuts are liked by many people better than the roasted ones.

Uncooked wheat may be used in several ways. Many children know how to chew a spoonful of wheat up until the starch is all dissolved by the saliva, leaving the pure gluten as a gum which can be chewed all day, as the alkaline saliva has no digestive effect on it. This kind of gum has the merit of being really nutritious, and will stimulate the flow of saliva and assist stomach digestion as much as any of the "pepsin" compounds with indigestible chicle which are so extensively sold.

Hard Northern wheat is the best kind, and it may be made into a very palatable dish by soaking for 36 hours, changing the water two or three times to prevent it souring. It is then so swollen and soft that it can readily be eaten with cream, or honey. It may be warmed before using. This is a sovereign remedy for constipation. Rye can be used in the same way and is generally regarded as more laxative than any other cereal. Flake rye, or rolled rye is a good breakfast food, but most people prefer to cook it about fifteen minutes.

Wheat may be parched in a hot skillet until the grains puff up and become crisp and palatable. This is called spargo, and may be chewed or ground up and eaten with cream or hot milk. Corn spargo can be made from dried sweet corn parched in the same way. Spargo is the principal diet of some of the Sicilian peasants.

Nuts are a good food, but rather concentrated. Most of them are improved by soaking in water for a day, before cracking.

Milk and sour milk are prepared in various ways to render them more easily digestible. It is well, in this connection, to remember that it is possible to give the stomach too little to do, and that a continuous feeding of pre-digested foods may weaken that organ.

I never advise the use of sweet milk as a drink with meals, but certainly buttermilk and sour milk can be used, in moderation, with good effect. It is difficult to get good buttermilk, and the so-called buttermilk tablets, pretending to be cultures of strange and beneficial germs from foreign lands, are mostly humbugs.

The best way to prepare buttermilk, or rather, sour milk, is to take the fresh, warm milk, and put it away in a glass fruit jar, in a warm place. If the weather is cool, or the drink wanted soon, place a teaspoonful of sour milk in the jar, as a starter. In about 24 hours the milk will be coagulated, or "clabbered."

Now empty the contents of the jar into a bowl and beat up with an eggbeater, until liquid and frothy, when it is ready to drink.

Junket tablets to coagulate sweet milk can be bought at drug stores, and many useful recipes come with each little box. With their aid most delicious and digestible dishes can be made from milk and eggs without cooking.

No matter what food is given the stomach, it makes an effort to digest it, and nearly always succeeds with any simple, natural food; but combinations of various classes of foods render the stomach's work more difficult. Professor Pawlow's researches have thrown great light on the subject of gastric and pancreatic digestive secretions. He showed that each kind of food ingested was instrumental in causing the secretion of digestive juices especially adapted to that food, but when several kinds of food were taken at the same time, a stomach had to be in very good order to make gastric juice suited to all.

To this fact, no doubt, is due the origin of numerous mono-diet cures.

Perhaps one of the oldest of these is the so-called grape cure, whch has been and is still used in various parts of the world. The method of giving this treatment varies in different places. Usually the patients go to the vineyard and live out doors while eating several pounds of grapes daily. As high as ten or twelve pounds can be taken, but the usual daily allowance is two to four pounds. In some cases a small amount of coarse bread is added to the fruit. The diet is quite laxative, and has been considered excellent for digestive disturbances and consumption. In some places the custom is to swallow the seeds while in other localities this is not advised. The skins are usually, but not always, rejected. The soreness of the mouth frequently caused by the diet is relieved by rinsing the mouth with cold water containing a little bi-carbonate soda. A disadvantage of this cure is fact that it is only available at one season of the year.

Another remedy is the apple cure, which is now available at almost any time of the year.

A very good remedy for malarial poison, and, perhaps, other disorders, consists in living exclusively on tomatoes for weeks at a time. In my experience with this diet the tomatoes were cooked, but with little or no seasoning. I presume the fresh, uncooked fruit would be satisfactory. Cranberries are said to have a curative influence on rheumatism, and the use of uncooked vegetables, as carrots, for instance, is beneficial in skin disease. Recently an exclusive diet of cooked turnip tops has been tried in a large Eastern hospital, for digestive troubles, and with excellent effect, it is said.

The meat diet, called the Salisbury treatment, was popular in this country twenty years ago, and may be valuable in certain cases, such as obesity. The meat used consisted of very slightly cooked, lean beef. I have known of patients taking as many as eighteen steaks daily, with a crust of bread, and usually hot water and orange juice to relieve the constipation. This method is dangerous in certain conditions and certainly has not the general application of the fruit cures.

None of the above named foods contain in themselves a well balanced ration for the human body, and sooner or later, the diet would have to be changed to prevent starvation.

Some writers claim that eggs are a perfect food, but certainly we cannot do with eggs what we can with milk alone. An egg contains splendid food material, but it is primarily intended for the propagation of the species, while milk is secreted purely for food purposes and by an animal closer to the human race than the bird family.

PULSE		PULSE		WEIGHT	
DATE	NO.	DATE	NO.	DATE	POUNDS

MEMORANDA.

DATE	NECK	ARM	FOREARM	CHEST		WAIST	HIPS	THIGH	CALF
				LARGE	SMALL				

INDEX

Ingram Content Group UK Ltd.
Milton Keynes UK
UKHW040811140423
420171UK00001B/16

9 781429 01